THE ASTRONAUTS

THE ASTRONAUTS
CANADA'S VOYAGEURS IN SPACE

LYDIA DOTTO

First published in 1993 by
Stoddart Publishing Co. Limited
34 Lesmill Road
Toronto, Canada
M3B 2T6
(416) 445-3333

Canadian Cataloguing in Publication Data

Dotto, Lydia, 1949-
The astronauts : Canada's voyageurs in space

Includes index.
ISBN 0-7737-2707-8

1. Astronauts — Canada.   2. Astronautics — Canada.   I. Title.

TL789.8.C2D68 1993   629.45′00922   C93-093154-8

Cover design by Brant Cowie/ArtPlus Limited
Typesetting by Tony Gordon Ltd.
Printed and bound in Canada

*Stoddart Publishing gratefully acknowledges the support of the
Canada Council, Ontario Ministry of Culture and
Communications, Ontario Arts Council, and Ontario Publishing
Centre in the development of writing and publishing in Canada.*

*This book is dedicated to
the memory of my father, August Dotto,
and of my cousin, Robert Raffin*

Contents

Acknowledgements

Many people made significant contributions to this book. I would especially like to thank the Canadian astronauts: Roberta Bondar, Marc Garneau, Chris Hadfield, Steve MacLean, Mike McKay, Ken Money, Julie Payette, Bob Thirsk, Bjarni Tryggvason, and Dave Williams. Without their generous cooperation, it would have been impossible for me to write this book. I would also like to thank Rob Stewart, Mac Evans, Bruce Aikenhead, Karl Doetsch, Parvez Kumar, and Alan Mortimer.

Lynn O'Keefe, the public relations officer for the Canadian astronaut program, was extremely helpful in arranging interviews and providing me with written material and photographs.

The research for this book was funded in part by a grant from the Government of Canada: Science Culture Canada Program.

I would also like to thank my editors, Greg Ioannou and Elsha Leventis.

Introduction

The *Challenger* accident in 1986 did more than delay the two shuttle missions that Canadian astronauts were originally scheduled to fly. It fundamentally changed the nature of the Canadian astronaut program, and the reverberations are still being felt nearly seven years after the accident, even though the shuttle is flying again and with great frequency.

The long post-*Challenger* delay forced the original six astronauts into some serious soul-searching. After jubilantly claiming the prize that more than 4000 people wanted — a place in Canada's astronaut corps — they abruptly came face-to-face with the downside: years of delay, continuing uncertainty whether they would ever fly in space, and anxiety about the potential derailing of their scientific careers. It forced them to redefine their role, to find worthwhile and satisfying work on earth while never losing sight of the ultimate objective: flying in space.

Even after shuttle flights resumed, Canadian astronauts were faced with the hard reality that they will likely never have the same opportunities for shuttle flights as their American counterparts. And their long-term prospects for missions aboard the international space station *Freedom* have been affected by the troubles that have beset that program in the past decade: a slipping schedule, a reduced crew size, perennial financial problems, and vocal opposition from critics of the manned space program.

But there has been good news as well — a lot of it in 1992. In January, Roberta Bondar was aboard the shuttle *Discovery*, working on behalf of more than 200 scientists around the world. The flight provided Canadian astronauts with invaluable experience — not only in space but also on the ground — that will help them prepare for international operations

1

aboard the space station. And Bondar's flight was only the beginning of a very busy year: in June, four new recruits came on board; in August, two Canadians were seconded to NASA for advanced training, and in October, Steve MacLean flew a shuttle flight on which he tested a machine vision system designed to improve the capabilities of the Canadian-built manipulator arms for the shuttle and the space station.

During the post-*Challenger* delay, a closer relationship was forged between the Canadian space science and engineering community and the astronaut program. Canadian researchers gained experience in microgravity science and developed a number of experiments that were performed by Bondar and MacLean on their flights. And work continued on two major research thrusts, space robotics and space physiology, that are essential and complementary components of humanity's move into space.

There is, of course, the question whether humanity will have any kind of a future in space. Serious social, economic, and environmental problems here on earth are growing more urgent, and if they remain unsolved, they may well preclude dreams of the sort envisioned by Isaac Asimov and Gene Roddenberry. However, stopping the space program would not solve these earth-bound problems. Despite what the critics say, the amount of money spent on the space program is relatively small, far smaller than what is spent on social programs such as health care and education.

Furthermore, little of the money spent on space exploration literally burns up on the launch pad; most of it goes into jobs and research here on earth. Of course, it could be used to provide jobs and support research in other fields, but the same could be said of just about any other endeavour. Exploring space is an "extra," like the arts and like many other scientific and technological projects, that is, on balance, worthy of the investment of some of our time, some of our intelligence, and yes, even some of our money.

Why we find it worthy is a complex question. The simplistic notion that looking beyond the horizon and seeking new challenges is inherent to human nature is only part of the answer. Yet, in these hard times, it's the one most often under attack. In a recent commentary, *Globe and Mail* science writer Stephen Strauss argued that even though ants and cats also have exploring natures, no one suggests that "if bugs and kitties are not on the next *Enterprise* to Betelgeuse their true natures will forever remain unrealized." This is true, but hardly the issue. Perhaps if bugs and kitties had larger brains, opposable thumbs, hammers, and some billions of dollars, they, too, would choose to go into space, driven by their exploring natures. And even if they would not, why does it necessarily follow that we should not? Should we arbitrarily dismiss the human impulse to "push the envelope" as completely unworthy of expression?

But this book is *not* about whether human beings should or shouldn't be going into space. This question is a subject of never-ending but usually tedious and fruitless debate, since the debaters typically are immovably entrenched in one of two opposing quasi-religious camps. One camp believes that the space program is among humanity's greatest achievements and most noble endeavours, and that it has and will continue to produce worthwhile benefits on earth, not the least of which is giving expression to the human spirit. The other camp believes that it's a big waste of money that should be spent on "more important" things on earth. (Between these warring factions is a subset of people, mostly scientists, who approve of unmanned space exploration as a legitimate scientific enterprise, but who oppose the manned program as too expensive, dangerous, and scientifically unproductive.)

This book is not an effort to tempt people out of whichever camp they've settled into. Instead, it is a book about some of the people who go into space and why they would ever want to — about what they go through to get there and what they discover when they do.

1

The New Astronauts

Since Yuri Gagarin's flight in 1961, when the first human was sent into space, millions of children have grown up knowing that space exploration is no longer science fiction, but rather an entirely feasible enterprise. For those of you who are about to join the Canadian astronaut program, the next 10 years promise to be the most exciting ever.

— BOB THIRSK

It was 2:01 on a Saturday afternoon in June 1992. Dr. Dafydd (David) Williams, staff emergency physician at Toronto's Sunnybrook Health Science Centre, was preparing to give a lecture on emergency medicine to a group of nurses when his pager went off. He'd been expecting the call, so he went over to the phone near the classroom door to take it. After a brief, noncommittal conversation, he hung up and, without commenting on the interruption, began his lecture.

No one else in the room knew he'd just been informed that he'd been selected from among 5330 applicants as one of four new Canadian astronauts.

"About 25 people were standing around watching me answer this phone call," Williams remembered. "They were saying, 'Who is he talking to?' I was trying not to look excited because I was not supposed to let anybody know this was happening. But inside I was jumping up and down."

He admits he had not been expecting good news because of

4

the lateness of the call. He'd been told that the 20 finalists, who'd spent the last week of May in Ottawa being poked, prodded, and psychoanalyzed, would be called between 1 and 3 p.m. on June 6, and he'd assumed the successful candidates would be called either first or last. "Between 1:00 and 1:05, I was a bit distracted. Then, because I was teaching, I got back into the routine of things." When the pager went off around two, his first thought was, That's it, I'm history. "I was actually hoping I wouldn't hear until about five to three. I thought they were either going to call the four right away or at the end, so I didn't want to hear anything at two." But the delay had no ominous meaning — it occurred simply because Arline Marchand, head of human resources for the Canadian Space Agency (CSA), had received some incoming calls from candidates who thought they'd be tied up and wouldn't be able to take a call from her.

Williams found it understandably difficult to go back to work and pretend nothing had happened, but he did manage a quick call to his wife, Cathy Fraser, to give her the good news. "She's been behind me 100 percent," he said. In fact, it was Fraser, a commercial airline pilot with Air Canada, who'd urged her husband to go after the astronaut job even before the competition was announced. He'd been interested in the idea since Canada had selected its first astronauts in 1983, but what directly precipitated his application was hearing a NASA astronaut talking about her career when he went with Fraser to a conference of The Ninety-Nines, an international organization of women pilots. "My wife was saying, 'Gee, Dave, you should do that.' I'm saying, 'Oh, no, they'd never take me.' But she knew that I'd wanted it from 1983 on, so when we came back, I said, 'I'll send in my CV and see what happens.'" That was in August 1991. Williams called the space agency every few months and was told it was planning to hire new astronauts but still wasn't sure when. Finally, after CSA ran an ad calling for new astronauts in January 1992, he resubmitted his CV and, along with 5329 others, officially entered the race.

Even then, he nearly missed his opportunity because he and his wife won a trip to Colorado. "My wife enters every contest known to mankind," he said. "We have never won a thing, but we won this ski trip to Vail. So I sent in my CV and away we go to Vail." When he arrived back, he found he'd made the first cut, and CSA had sent him a bulky package demanding answers to hundreds of questions. The deadline was the next day. "We got back early in the morning on Thursday, and I spent the whole day Thursday doing it — it took close to eight hours — and got it in the mail by five o'clock Thursday evening. It had to be postmarked no later than Friday morning. So I managed to sneak in under the wire."

Three other people who'd been anxiously waiting by the phone that Saturday afternoon also got good news. They were a far-flung group: Julie Payette, a computer engineer who worked for Bell-Northern Research Ltd. in Montreal, was waiting for the call at the Hamilton home of her boyfriend, François Brissette, who was completing his Ph.D. at McMaster University; Major Chris Hadfield, a Canadian Forces exchange officer, was in Lexington Park, Maryland, where he was a test pilot at the U.S. Naval Air Test Center; and Robert Stewart, a University of Calgary geophysicist, was attending a scientific conference in Paris.

They'd all made the assumption that Marchand would call the successful candidates first, which made the period leading up to the appointed hour one of high anxiety. Chris Hadfield eased his tension by spending the early part of the day waterskiing with friends. "I figured, I'm not going to sit around the house until one o'clock sweating bullets. My wife said she couldn't go waterskiing; she was too nervous. So *she* sat around the house and sweated bullets all morning. I came back around quarter to one, and we were sitting around the kitchen waiting. I figured that if they hadn't called by about 1:30 or 1:45 then I'm so far down the list that I'm out of there. The phone rang at 1:09, so I figured, 'All right, I'm in.'" As for his wife, Helene, "as soon as she saw me smile and give her a thumbs up, she was jumping up and down and doing cartwheels. She's more demonstrative than I am."

His wife had known about his ambitions for a long time. The two started dating in their teens, and "I told her then that some day this is what I'd like to do. So this was not a surprise, and she's very supportive." Nor did the prospect of pulling up stakes faze the couple, who are used to the nomadic life of a military family. "We've lived in eight houses in the last ten years," Hadfield said, adding wryly, "This is a chance for us to settle down." As for his three children, "everybody they know is a test pilot," and they also know some astronauts so the fact that their dad is going to be one is not a big deal.

Meanwhile, in Hamilton, Julie Payette received her call around 1:15. "It was a long hour before that," she admitted. After five months of jumping hurdles, never quite believing she'd made it to the next round, she was more than ready to have the matter settled one way or the other. "It was the uncertainty that was hard to live with. This meant a big change — I was either branching into this other career objective I had, or I was completely changing my life and becoming an astronaut. I needed to know, and as soon as I knew, everything was very calm."

With the pressure off, she was not above playing a joke on her family in Montreal. Brissette, who placed the call, told her sister that Payette had locked herself in the bedroom and wouldn't come out. As Payette recounted it, "he said, 'The phone call came in negative and she doesn't want to talk to me or anything.' Then he says, 'She's coming out; she doesn't look good.' So I get to the phone and [my sister] says, 'How are you doing? I know you're feeling bad . . .' and I said, 'I'm an astronaut!' She just yelled into the phone. We are pulling her leg all the time."

In Paris, it was early evening when Rob Stewart got his call. During the day, he'd diverted himself with a trip to Versailles, but the train was late getting back to Paris so "I was madly scrambling to get back to the hotel room." Arriving just before 7:00 p.m., he then spent what seemed like a very long 20 minutes pacing around the room until the call came in. That evening, he celebrated with a colleague who'd recommended him for the astronaut program at a restaurant on the Champs Elysée.

These calls were the culmination of an intense and gruelling five-month process that, like the first astronaut selection in 1983, had started with a help-wanted ad. The ad, which ran in January 1992 in major Canadian newspapers and two U.S. space publications, announced that the Canadian Space Agency (CSA) was seeking astronauts and offering "exciting career opportunities," including the possibility of conducting scientific experiments aboard the space shuttle and the international space station *Freedom*, as well as operating remote manipulators and performing space walks. Candidates had to be Canadian citizens and possess at least a bachelor's degree in engineering, the physical or biological sciences, medicine, or mathematics. An advanced degree or at least three years of professional experience was also required.

There were several other important requirements. Because astronauts are always in demand for media interviews, public speaking, and educational activities, they must have good communications and presentation skills. Because they have to work with scientists, engineers, and astronauts from many other countries, they must possess good interpersonal skills and be capable of teamwork. And they have to be in top shape physically and mentally. The ad noted that "because astronauts have to work under difficult physical and psychological conditions, applicants must be willing to provide medical and personal information and to undergo very demanding physical and psychological examinations . . . in accordance with the requirements of the Canadian Space Agency, NASA, and the other international partners of the Space Station *Freedom* Program."

The agency had started to consider selecting a new group of astronauts more than a year before, for two major reasons. One was that the U.S. space agency, NASA, had offered to accept two Canadians into its training program for "mission specialists" — astronauts with operational responsibilities on shuttle missions, whose duties include running and, if necessary, fixing equipment, doing space walks (known as extravehicular activity, or EVA), and operating devices such as the Canadarm, the shuttle's Canadian-built remote manipulator arm. Up to that point, only U.S. astronauts could be mission specialists. The first group of Canadian astronauts, selected in 1983, were known as "payload specialists" — astronauts restricted to doing scientific experiments on behalf of CSA and

Canadian researchers. Payload specialists have no operational responsibilities and, in the past, have tended to be viewed by NASA more as passengers than as astronauts (although this attitude now appears to be changing).

By the end of 1990, CSA had accepted NASA's invitation: two Canadian astronauts would be sent to the Johnson Space Center in Houston during the summer of 1992 to begin four to five years of training, culminating in a shuttle flight for at least one and possibly both. Although it was expected they would ultimately return to give the Canadian program the benefit of their new expertise, they would be largely unavailable to CSA during their stint in Houston, so Canadian space officials felt that it was time to inject some new blood into the Canadian astronaut program.

The second reason for a new recruitment drive was that CSA felt the time had come to start preparing Canadians for duties aboard the international space station *Freedom*, which was being developed jointly by the United States, Canada, Japan, and Europe and was then expected to be in operation sometime around the turn of the century. The future of the Canadian astronaut program is closely linked to Canada's involvement in the space station; without the prospect of regular station missions, it's questionable whether Canada would continue to support a full-scale astronaut program. However, no one knew exactly when Canadians (or anyone else) would be flying those missions, since the program was perennially in financial trouble in the U.S., so deciding just how many new people to bring on board was a tough call for CSA, requiring about equal measures of optimism, crystal-ball gazing, and a hard-nosed assessment of the financial realities of the space program over the next decade.

There's no doubt, however, that flying a space mission is *the* bottom line in the astronaut biz. While there's plenty of interesting work to do on the ground, the lure of floating high above the earth is ultimately the carrot that attracts prime talent as astronaut candidates. But whereas the first group of astronauts had come into the program expecting to fly two shuttle missions over a three-year period, those in the second group were warned that it could take much longer for any of them to get a flight, that mission specialists might have to wait up to four to five years, and those assigned to space station flights could not expect to fly until around the end of the decade. Payload specialists might fly sooner, but this would depend on "flight opportunities that are yet to be negotiated," said Bruce Aikenhead, then director general of the Canadian astronaut program. "So it will be a long wait for some, but potentially there will be earlier flight opportunities. There are no guarantees; there is, however, a very high probability that Canadian astronauts will fly."

For CSA, this question of numbers was further complicated by the fact that only one of the original six astronauts chosen in 1983 — Marc Garneau — had managed to get a flight before the *Challenger* accident put the shuttle program on hold for several years. Two others — Roberta Bondar and Steve MacLean — were scheduled to fly in 1992. At least one more Canadian astronaut was guaranteed a mission in the mid-1990s, but other than that, there were no definite flight commitments. Adding to the uncertainty was the fact that the space station program was enduring the usual cutbacks and threats of outright cancellation from the U.S. Congress. And the Russian program, which had been forging dramatically ahead of the U.S. program, abruptly ground almost to a halt as political revolution and desperate economic problems engulfed the former Soviet Union. Thus, it was unclear how robust the manned space program would be during the 1990s, and this uncertainty was one of the main reasons why only four new astronauts were chosen, even though CSA officials had been saying virtually to the end of the competition that they would select up to six.

In fact, several of the existing astronauts had argued strongly that a lower number was more appropriate, given the uncertainties about flight opportunities. "We had a fair bit of input into narrowing down the number," said Bjarni Tryggvason. "We recommended it should be three, and it took a long time for them to back off from six to four."

The number of new astronauts was not the only issue to be considered. CSA also had to decide what kind of people it was looking for in terms of skills and personality traits. Because only payload specialists were selected during the first competition, there had been a strong focus on scientific credentials, particularly in the two major areas of Canadian interest, space robotics and space physiology. While these would still be important in the second round, additional criteria were also important for people expected to qualify as mission specialists and space station astronauts. There would, for example, be an increased premium on the ability to operate complex equipment — "eye-hand coordination, quick responses, and that kind of thing," said Garneau, who had examined the benefits of mission specialist training for CSA. "The requirements for mission specialists may put an emphasis on operational skills and not quite so much on research background. The person has to have some of each; it's a question of emphasis." Both mission specialists and payload specialists (and their equivalents on the space station) "must possess many common aptitudes and skills," he said. Both must be able to "get the job done" and "fix things."

Aikenhead emphasized that this new approach didn't mean that scientific research skills won't continue to be needed. "It will be important for agencies such as ours to continue to have payload specialists." Flexibility is the key ingredient; the agency was looking for "people whose academic backgrounds and experience show that they can cope with the lab work and science, and hands-on operations, [people] who have general capabilities to act on behalf of the engineering and scientific communities as well as worry about space station operations. We're really trying to find people who have that versatility."

CSA determined as early as the spring of 1991 that a new selection campaign was necessary, and they were anxious to get it underway quickly, since they'd decided that the new recruits should be selected in time to be eligible for assignment to the mission specialist training program scheduled to begin in the summer of 1992. But the CSA was a relatively new agency, formed in 1989, and it lacked the administrative structure to handle the expected deluge of applications. In 1983, the National Research Council, which was then managing the Canadian astronaut program, had anticipated about 1000

applications and got 4300. This time, Aikenhead said, they
assumed they'd get perhaps 2000 to 3000. "We were conjectur-
ing — has the novelty of space worn off?" Obviously not —
they got 5330 applications.

During the several months it took CSA to find an independent contractor to manage the campaign, the agency tried to
keep it quiet that they were in the market for new astronauts
to avoid being swamped by calls and letters before they were
ready to deal with them. In fact, their ad for the contractor
coyly stated that the job was to help CSA find "scientists and
engineers" for a variety of challenging tasks. "A clever person
could perhaps read between the lines," Aikenhead acknowledged. But the statement was literally the truth, if perhaps not
all of the truth — CSA did need more scientists and engineers,
and Aikenhead said CSA expected many applicants "who
might not make it as an astronaut, but who ought to be able
to fill useful positions within the agency, so we could establish
a data base of extremely well-qualified people doing aerospace work. That certainly has been a by-product."

All this sleight of hand went for nothing, however, because
the beans were spilled in late June at a media briefing for
Roberta Bondar's flight when, to the bemusement of several
CSA officials, Larkin Kerwin, then CSA president, and science
minister William Winegard spoke openly of the agency's intention to recruit new astronauts. These statements were of
course picked up by the media, and the phones started ringing
in the astronaut office the very next day, with calls coming in
from around the world. "We were deluged," said Aikenhead.
"One of the human resources people got a call. She said 'Could
you speak up, we seem to have a poor line. I can call you back.'
And the fellow said, 'No, no, I'm calling from Karachi.' This was
the following day. It was pretty intense here for about three
weeks." It was also a portent of things to come. In fact, the ad
calling for new astronauts appeared only once because, as
Aikenhead put it, CSA knew "it was not going to be necessary
to run this thing day after day."

CSA chose a Montreal company, Selectra Consulting Group
International, to manage the campaign, and when it finally got
underway in January 1992, it followed much the same pattern

as the first. In response to the ad, the applications came pouring in from all over the country and from Canadians abroad. Some came from as far away as Australia and New Zealand.

In the first competition, applicants ranged in age from 6 to 73, and many did not have suitable qualifications; in the second, CSA hoped the published list of qualifications would discourage inappropriate applications, but there was no way to forestall them completely. "Experience shows that you can have the most precisely worded set of criteria in the advertisement, and you're still going to get people who are woefully unsuited who will apply anyway," Aikenhead said. "I think of some of the correspondence we've had — paraplegics, for example, who say, 'I have read that legs aren't necessary in zero gravity, therefore I want to be considered as a payload specialist.' You have to write a kind reply to it. There will be letters from kids. There will be a lot of people who'll say, 'I never went to university, but I'm as good as the next guy.'"

A strictly objective scoring system was used to rank the applicants based on the information they sent in about their academic background and work experience. "It was a pure paper screen," said Selectra president Alan Davis. "If you had a bachelor's degree, you got a certain number of points; if you had a master's degree, you got a certain number of points; if you had a Ph.D., you got a certain number of points. If you had cross disciplines, you got extra points. If the relevancy of your degree was high, you got more points. We also weighted experience, and came up with a score." Selectra then ranked the top applicants, being sure to get a good mix of the professional disciplines that were of major interest (for example, engineering and life sciences), and presented CSA with about 550 people for further consideration.

CSA made the first cut to 370 applicants, and these people were sent a specially designed six-page application form, along with questionnaires covering medical, psychological, and security matters. Because they would potentially have operational responsibilities on board the space shuttle or the space station, the candidates in this competition had to meet a higher medical standard than those in the first recruitment. And because space station missions will last longer than shuttle flights — from three to six months in the early years, compared with less than two weeks for most shuttle flights — there was much greater emphasis on evaluating the maturity and psychological stability of the candidates. In fact, one question the members of the selection panel kept asking themselves as they interviewed and assessed the hopefuls was, Would I want to spend three months on a space station with this person? "You are really looking at the person twenty-four hours a day, not just through eight hours on the job," said Aikenhead. "What is this person really like, day in and day out?"

The medical forms returned in the second round were sent to Dr. Gary Gray of the Defence and Civil Institute of Environmental Medicine in Toronto, who evaluated them according to NASA's standards for mission specialists. The psychological questionnaire was passed on to Dr. Daniel Sommer of Sommer et Belanger, a Montreal psychological consulting firm. The security forms were routed to CSA's security office, and Selectra evaluated the application forms. The medical was a show-stopper for many people, said Alan Davis. "A lot of people failed on vision. We were measuring people against mission specialist requirements; the eyesight requirement is very acute, and a lot of people didn't make it." This did not automatically eliminate people who wear glasses or contact lenses, but a candidate's visual acuity without them was still very important; if astronauts were to lose their glasses or contact lenses during a critical manoeuvre, they should still be able to get the job done.

This round cut the hopefuls to about 100, a manageable number for the first set of face-to-face interviews. Selectra dispatched a recruiter and a psychologist to five Canadian cities and to Washington, D.C., where they talked to candidates and administered another psychological test. Davis said the purpose was not "just to get information from the applicants, it was also to give information. We spent a lot of time telling them what the job is all about — that a lot of travelling is involved, and it's very disruptive to family life. We wanted to lay it out so the people who passed that stage would be genuinely interested, [so] they could effectively screen themselves out if they weren't interested." It was particularly important to make it clear to the candidates that becoming an astronaut did not automatically guarantee them a space flight. "A lot of people have this feeling they call you up and you go down to NASA and get on the next shuttle when it's your turn," Davis said.

And then there were 50. At this point, a seven-member CSA panel, which included astronauts Bob Thirsk and Marc Garneau, travelled to Montreal, Toronto, and Calgary to conduct further interviews and assess the candidates in person. Thirsk, who had been a member of an international group that defined the selection criteria and training requirements for space station astronauts, found interviewing the candidates challenging because they were so diverse in their qualities and skills, and the panel had only a short time to assess each one. "Our goal was, in one hour, to find out who these people were. So you have to ask some pretty probing questions. There were times when the applicants became emotional, in a positive sense; they gave wonderful things about themselves or memorable experiences they've had. There were also instances where we probed so much that we hit vulnerable areas, and some cried right in front of us."

But he added that the panel did try to set the interviewees at ease as much as they could. "I would take extra time to explain the question so the applicant knew exactly what you were asking. And I would always try to smile, be cheerful, to throw a joke in once in a while. There was give and take, with spontaneous questions in addition to the mandatory questions, and it got to be like a conversation, the ideal type of environment where you can really get a good impression of who an individual is." This was a far cry from the situation he'd encountered in 1983 when he was sitting on the other side of the table. "I know the selection panel then had instructions not to laugh at our jokes." (In fact, Steve MacLean once described their interrogators as a group of "stonefaced interviewers who were programmed not to react to us.")

The 50 applicants did more psychological tests and for the first time were physically examined at a Canadian Forces medical facility. Again, some candidates failed to meet the stringent medical standards, often because of eyesight. "For both

the recruiters and the applicants, this was the hardest area —
being screened out for medical reasons," said Davis. "We had
a lot of disappointed people all through this process."

The applicants were also required to participate in their first
media briefing. Asked why they'd applied for the program,
several candidates commented that becoming an astronaut
had been a childhood dream that had set them on their scientific career paths. "I think a lot of us share the feeling that being
an astronaut is a way to touch the future," said Paul Cooper, a
Canadian researcher working at Northwestern University in
Illinois. John Boyce, a University of British Columbia scientist
who had been involved as a ground-based researcher in several shuttle experiments, commented, "I'm tired of looking at
all the pictures that are coming down [from space]. I want to
see it firsthand."

For the selection panel, the media conference served as a
test of the candidates' abilities to handle the public relations
demands of the job. Certainly, Payette was one who assumed
that was the purpose of the exercise: "I knew the press conference was to see how we would behave. There was no point in
presenting the 50 semi-finalists otherwise."

From the group of 50, the CSA panel selected 20 finalists to
invite to Ottawa at the end of May for yet another round of
evaluation and testing. The hectic week was packed with interviews and presentations for the selection panel, another
media conference, and the never-ending medical and psychological tests. Doctors at the National Defence Medical Centre
put the candidates through an even more thorough medical
examination. "They tested everything imaginable," said
Davis. "They've got all the latest equipment, so they were able
to test in more detail. They did about 20 tests each, some of
which took five minutes and others took three hours. The
degree of testing got harder and harder, and there were some
of the 20 who were cut medically."

"That was a heartbreak for them," said Aikenhead. "It was a
bit of a heartbreak for us as well, because they were all very
qualified. We just hated to see them ruled out on that ground."
In retrospect, he believes that the medical screening should
have been even more rigorous at the very beginning to avoid

these disappointing eliminations so far along in the selection process.

As for the psychological tests, "they ran us through things that just drive you crazy," Hadfield punned. "Some of the questions are just ridiculous. One was: 'Do you lie?' Well, what a useless question to ask. What if I say, 'Okay, you caught me, I lie'?"

Another question offered a list of disparate qualities and asked, "Which one of these is most like you, and which one of these is least like you?" But they would be completely unrelated issues, and Payette said that in some cases, none of the qualities applied, and in others they all did. "I've never answered so many questions in my entire life. They were a real pain."

"And very long. Hundreds of questions," added Hadfield, who found them more tedious than intrusive. "When I was going through the tests, I was thinking it would be very easy to fool these people; you just have to keep giving the obvious answer that you think they're going to want. But then again, I look at the people they chose, and I think they've managed to find people who get along pretty well and without any significant prejudices that are going to be a problem on a team. So I think the process worked fine."

Aikenhead said the CSA selection panel was impressed with the insights provided by the psychological profiles. "We would see CVs of people we knew. The psychologist knew these people only as numbers and . . . in some instances, it was uncanny to hear somebody described so accurately." At times, psychological factors could tip the scales in one direction or the other, he said. "We might have a case where somebody has a marginal CV, but the personality factors indicated that this person is very interesting and dynamic, so that might nudge it just a little. There were some instances where somebody . . . has a very interesting CV but . . . could be a loose cannon, a guy who, according to his questionnaire, will break all the rules."

The intense analysis of personality was the most marked change from the first recruitment campaign, which sought astronauts only for short-term shuttle missions. The second group was required to qualify for space station missions lasting months and involving multinational and mixed-gender crews.

All the countries involved in the space station program are concerned about issues of crew compatibility and want to weed out people with abrasive personalities or sexist or racist biases from the outset.

Other qualities in demand were high levels of motivation and discipline, the ability to lead without being overbearing, and maturity in handling stress and failure. "You're up there, and of course you want to succeed in the experiments, but what if you don't?" said Arline Marchand. "Can they withstand the pressure, thinking they may not succeed and something might go wrong?"

Psychologist Daniel Sommer said that after discussion with officials at CSA and NASA, a psychological profile of the desirable personality type emerged. "You want people you feel you can rely on. You don't want whiners. You want people who take things in stride and share success. People with strong egos, but low ego needs — who know when to be assertive and when they have to yield." What's needed, he said, is a "flexible type of leadership," people who "can be leaders and also can be followers. Some people have to have their way all the time and [they] may be hard to live with. On space station, if you're stuck with that person for six months, they can grate on your nerves."

Sommer said they were also looking for people with high ethical standards and who demonstrated conscientiousness and qualities of benevolence — a willingness to help others — and, of course, the ability to handle stress. "Tolerance to stress is paramount. People have to keep their cool, to accept ambiguity, to accept some inherent danger in a mission, and delays and frustrations." Intelligence and intellectual curiosity were also important, but more than just being smart, the successful candidates had to be quick. "You need people who can learn on the fly, especially highly technical stuff. Some people can have all kinds of degrees, but they're not extremely swift; they do it by dint of effort and time."

The candidates were also assessed on a scale of introvertedness and extrovertedness. Scientists are often introverted, Sommer said. But in the Canadian program, much more than in the U.S. program, astronauts are expected to play a major role as advocates of the space program. So CSA wanted people who could comfortably deal with media, politicians, the public, and school children, but at the same time, "we don't want show-offs, people whose heads get swollen by the fact that they're astronauts," said Sommer. (Certainly, the first impression that one gets of the four successful candidates is that they are self-confident without being arrogant. And, while they seemed a bit overwhelmed by the sudden and intense media attention — just as the first group had been — they handled it with aplomb.)

One of the most important questions candidates were asked was to list their strongest and weakest qualities. (Dave Williams wryly noted that "with the five strongest ones, I got as far as two and they interrupted me. With the five weakest, of course they wanted all five.") Sommer said this question has "tremendous value. It separates people who are mature and have self-knowledge from those who do not have self-insight and do not accept criticism." He was not seduced by answers like "I work too hard," which he described as an attempt to claim "a quality disguised as a shortcoming." He was interested in how people overcame their shortcomings, so he asked them to describe incidents involving these problems and "how they dealt with it, and the consequences on friends and family and co-workers."

Sommer did not interview the candidates' families, but he did ask questions about their personal lives. He explained that they were not required to answer and that the answers would remain confidential, but "I had to ask questions about family, relationships, parents, and so on. I wanted to know what kind of upbringing they had, their values." He was also interested in knowing if spouses were supportive, to assess whether "this kind of move is going to dissolve the marriage or create strain." He encountered "very little resistance" to these questions and said the information would be destroyed after two years.

Even the leisure and sports activities of the candidates were examined. Active participation in sports could indicate that a candidate had a high level of physical fitness, leadership qualities, and the ability to work with a team. Water sports were of particular interest, because survival training requires astronauts to endure a lot of dunking, and practising for space walks requires working in a large water tank. "There were people who would have a real problem [with] the underwater work because they didn't like swimming," noted Aikenhead. Community work was also of interest because these were "activities that reveal not only things like leadership, but concern for others," he said. "We found some interesting cases where people, in addition to all the other things they'd been doing, had been very active in . . . working with the aged or problem kids or the infirm or what have you. Quite a number of them had done voluntary work long before they ever thought about applying for an astronaut position."

Alan Davis said such activities are "a help in assessing what these people are like. It's an indicator of the right sort of person, if they have a balanced lifestyle . . . a high level of academic or work achievement plus a whole bunch of other things, sports, community work. What you typically find is that people with a broad range of interests and a high level of achievement tend to be the ones with a high level of communications skills and very high levels of motivation. People with very narrow interests tend not to have the interpersonal and communications skills needed for a job like this."

Everyone agreed that the 20 finalists were an impressive bunch, but highly varied in their background and skills. "It was an interesting group of people," said Hadfield. "I wish we'd had a chance to all go camping together." When he was asked after he'd been selected whether the diversity of people in the final round had raised questions in their minds about exactly what CSA was looking for in an astronaut, he laughed and said, "Still don't know." He admitted it was difficult to avoid second-guessing and trying to "bring out personality traits that you think are what they want out of an astronaut. That's a natural thing to do." But, in the end, all they could do was be themselves and hope that fit the bill. Williams said, "I went in with the attitude [that] I can't change my background, [and] I can't change my personality. If it will be sufficient to allow me to be accepted, great. If not, well, I tried my best."

One thing the candidates could do nothing about were factors unrelated to scientific or technical qualifications or to personality, so-called "secondary" factors such as gender, race, ethnic background, or regional origins. In the first competition, there was some discussion about the touchy issue of whether these factors would be part of the selection criteria. CSA officials, especially those directly responsible for finding people who can get a highly complex job done, have always steadfastly denied that secondary factors play any role. Prior to the second recruitment, Aikenhead said, "If I've got anything to say about it, there will be no slant towards regions, races, genders, or any of that. There's just no requirement for it, and I think those sensitivities, if there were any, are a closed chapter. We should be able to go strictly on the basis of academic and experience backgrounds and physical fitness."

Nevertheless, the issue usually crops up. For example, the temper of the times always makes gender representation an issue; during the first recruitment, CSA officials, while denying that there was a quota for women, acknowledged a certain relief at finding a candidate as well-qualified as Roberta Bondar, which allowed them to select a woman without "forcing the system."

Bondar was the only female finalist in 1983; there were two in 1992. Payette commented that "we were saying to each other, 'I hope at least one of us makes it.'" Asked why not both, she replied that she felt this was "completely out. When there are two [people] with some sort of characteristic within a group, the odds that both are chosen is very low." (With similar reasoning, Hadfield believed that only one of the two test pilots in the group of 20 would be chosen.) But Payette rejects the idea that she was chosen because she is a woman. "I didn't come into this group by the back door; I had the same testing as everybody else, so they had the same criteria to judge me. Once the selection was made, I am sure that it creates a balance that they're very happy with. But you want to believe they will take the four most adequate [people]. If there is no woman in those four, then so be it."

Together, Bondar and Payette represent about one-fifth of the astronaut corps. Similarly, Payette and Garneau make up a 20-percent francophone element. However, the astronaut group is not racially diverse, and most of its members have roots in Ontario and Quebec, although they have studied and worked all over the country, as well as in the United States, and therefore can be (and are) claimed by many communities. Indeed, newspapers in the cities where the astronauts were born immediately staked their claims to the new celebrities. "Latest Canadian astronaut likes Western Canadian label," announced the Saskatoon *Star-Phoenix* in a story about Dave Williams, who was born in the city, but left with his family before the age of two. The Sarnia *Observer* ran an article about Chris Hadfield with the headline, "City native named astronaut can't wait to get into space." As for Julie Payette, she was overwhelmed by the demands of the Quebec media. "People there feel things as a family. To have a hometown person making it on the English side — they just went crazy."

Age was a factor in the selection process because CSA was expecting these people to stay in the program for a decade or more. "If we chose somebody already in the mid-50s and asked them to stay for 15 years, it's a bit much," Aikenhead said before the selection began. "Mid-40s is about as high as we would want to go."

"Actually, the process looked after that [with] the people having to undergo all these medical examinations and fitness testing, but it was a young group," said Arline Marchand. "When we were at the 50 stage, the youngest was 25, the oldest 44, and the average was 38." Among the final four, the average age was down to the early 30s. The first group of astronauts had been slightly older on average, but at that time, CSA was looking for a short-term commitment.

There was a significant difference between the two recruitment programs on the question of bilingualism. In the first round, it was not required that all the successful candidates be bilingual at the outset; rather, it was stated that the astronaut group as a whole would have bilingual capability, which, in practical terms, meant that there would be at least one fluently bilingual astronaut who could carry out public relations duties in French. Initially, that capability was provided by Garneau, who was raised in a bilingual home. But the other astronauts were given language training, and most became proficient enough to give speeches or do media interviews in French.

In the second round, the bilingual requirement was stronger. Although it was not stated as a requirement in the ad, the 370 candidates who were sent the detailed application form after the first cut were asked to respond to essay questions in both their primary and a second language. ("I thought about using Welsh," said Williams, whose first name is officially spelled in the Welsh form. "But I don't read it, write it, or speak it [so] that would have been a real challenge.")

But qualified candidates were not rejected for being unilingual, since language instruction would continue to be offered within the astronaut program. "If we found somebody who was absolutely a front-runner but who was unilingual, we would slant the training program such that [they] become bilingual," said Aikenhead. "We do intend that there be some period set aside for language instruction, either to enhance proficiency in the other language or, if they're already totally fluent in both French and English, to learn a third language."

In the past, the bilingual requirement for Canadian astronauts was primarily intended to satisfy domestic considerations, such as the need to interact with francophone scientists, media, and the public in Canada. But the growing requirement to work with multinational crews and scientists is making multilingualism an increasingly important asset for all astronauts. Roberta Bondar's mission, which involved more than 200 researchers from all over the world, was a harbinger of things to come, both in the shuttle and the space station programs. English will continue to be the operating language in space — all space station crew members must be fluent in written and spoken English — but the participating countries all believe that multilingualism will help crews work together and understand each other better. "Not only does it have the potential of being a real advantage operationally, but it also gives a better entrée into becoming more acquainted with another culture," said Aikenhead. "We would prefer to have bilingual people who become multilingual — we would like to encourage somebody to study Japanese or Spanish or what have you."

One of the new astronauts has quite a head start in this respect: Julie Payette is fluent in English, French, and Italian and can also speak conversational Spanish and German. The others are all fluent in both French and English.

At the end of the last week of May, the members of the selection panel gathered to make a decision. It was tough. "We were talking about people who were very strong in just about every category you could imagine," said Aikenhead. "But some were stronger than others, so now you were starting to horse-trade back and forth." Each of the committee members ranked the candidates individually, and they ended up with "a small cluster at the top. Some were clear front-runners — because of their personality and experience they just tended to bubble up. And then there were others who were very close." In the end, they voted. Marchand said that unanimity was not necessary, but everyone on the panel had to feel comfortable with the final choices.

By the second week in June, CSA was finally ready to reveal its choices to the country. On June 9, at a media briefing in Ottawa, the agency presented the cream of the latest crop of aspirants, four young people who had accumulated an impressive list of credentials and accomplishments.

Julie Payette, at 28 the youngest of the four, had been a computer engineer who had been head of a research project at Bell-Northern Research Ltd. that involved developing new techniques for computers to use human language and recognize speech. Before that, she was a visiting scientist in the communications and computer science department of IBM's research laboratory in Zurich, Switzerland. She received engineering degrees from McGill University and the University of Toronto and was the spokesperson for the U of T's 1990 Women in Engineering Campaign. Payette is also active in sports — she participates in triathalon, skiing, racquet sports, and hiking — and is an accomplished musician and singer, having sung with the Orpheus Singers and the Montreal Symphony Orchestra Choir in Montreal, as well as professional choirs in Switzerland and Toronto. In addition, she volunteered her time to sing and play music at hospitals and retirement homes.

Her wide-ranging interests reflect an enthusiasm for new experiences. When asked at the media briefing what work she'd like to do if she was assigned to a space mission, she barely hesitated before answering, "Oh Lord, everything." She admitted to being a bit "flabbergasted" at having been selected as an astronaut, however. "I had wanted this since I was a teenager; I often say I wanted to become an astronaut like others wanted to become a ballerina or a fireman. It was something I had dreamed about, thought about, without really knowing what to do about it, [so] when I saw the ad there was no way I could pass it by." But she didn't go into the competition with high expectations of success — "I felt that maybe I was too young and didn't have so many diplomas, and I have only three years in research" — so she told few people she'd applied. "I really didn't say much about it. I'd just started a new job; I was happy with the job I had, and there was no point in starting to say I'm quitting for something else." But, to her amazement, she kept making the cut to the next round. "It was going fast . . . pressure increasing all the time. Every week we had something different to do, another test, another this, another that, and then you find yourself for a week in Ottawa. You meet these extraordinary people and you think, 'They are too good.' You are just living through this, feeling a bit numb, until the final phone call comes. You find yourself in the astronaut seat, and you didn't even see anything go by. What I felt the strongest was relief."

Like the others, Payette had to consider what being an astronaut would do to her personal life, and she discussed the question with her boyfriend before applying. The couple married shortly after Payette was selected, but their relationship, already a long-distance one while he completed his Ph.D. in Hamilton, continued as such for a while. At the time of her selection, Payette noted that "the space agency is relocating to Montreal next year, and that will coincide with his return to Montreal, so it will work out nicely. We're happy to have a home base in Montreal, because both our families and most of our friends are from there. But we both like to move, so if they send me down to Houston, I won't say no and he won't say no either. He'll be a university professor. He works in environment, water resources, things that are very important subjects everywhere in the world. So he knows he can find a job somewhere else."

*C*hris Hadfield, at 32, had earned degrees in mechanical engineering and aviation systems, had graduated from the U.S. Air Force pilot school at Edwards Air Force Base in California in 1988, winning the school's outstanding graduate award, and had been assigned as a Canadian exchange officer with the U.S. Naval Air Test Center, where he won the U.S. Navy's award for test pilot of the year in 1991–92. His duties included evaluating the handling qualities and performance of tactical jet aircraft and developing new procedures for recovering from out-of-control flight in the F/A-18 fighter. In an interview with CBC TV, he explained that this involves forcing the plane to go "completely out of control — it's tumbling end over end — and then we evaluate different ways to recover it. It's been pretty interesting." When the interviewer suggested that a shuttle flight sounds "a whole lot safer" by comparison, Hadfield commented, "I think they're both equally safe." But he did concede that "a lot more people have died testing airplanes than have in space missions."

Prior to his U.S. assignments, Hadfield was a fighter pilot at Canadian Forces Base Bagotville in Chicoutimi, Quebec. At the time of his selection, he had logged more than 2000 hours flying over 50 different types of aircraft. He has won numerous awards and honours and is a member of the Society of Experimental Test Pilots and of Mensa. His leisure pursuits include downhill and waterskiing, volleyball and racquetball, scuba diving, sailing, and horseback riding.

Hadfield comes from a family with flight in its blood — his two brothers and his father are airline pilots. He remembers holding the controls of an airplane for the first time at the age of three or four, and he got his pilot's licence on his 17th birthday. Naturally, his father wanted him to become an airline pilot, too, "because that's what our family does." But at the age of 10, he watched Neil Armstrong step out onto the surface of the moon, and he knew where he wanted to be. "I've been murmuring quietly since I was 10 years old that I wanted to be an astronaut, and it's just been getting louder recently," he told the CBC before being selected. This goal influenced his educational and career choices as he grew up. "The reason I went to college, the reason I got a master's degree, the reason I went to be a test pilot was to attempt to qualify. It wasn't that I was going to be disappointed if I didn't get to be an astronaut. I just thought, that's a good direction to head in, and I will have all the fun I can along the way. And as it was, I ended up having an interesting time. I would have been very happy to continue as a test pilot."

He didn't apply for the 1983 competition because he didn't feel he was qualified enough, but he contacted CSA as soon as he got wind of the second recruitment. Since he was about a year too early, he was told to watch for an announcement in the papers. Living in the U.S., he didn't have regular access to Canadian newspapers, so he asked friends in Canada to keep an eye out, and within days of the ad's publication, he received several faxed copies. Since he still "belonged" to the Canadian Forces, he checked with their representatives at the Canadian embassy to make sure it was all right for him to apply, and they gave him the green light.

"The timing couldn't have been better," he said. "I'd completed all the qualifications that I wanted to just in time." If an opportunity hadn't come up in Canada, he might have considered applying to NASA, but that would likely have meant becoming an American citizen. "I'd much rather do it for the country in which I was raised," he said. "I'd always hoped there'd be a Canadian program, [and] I'm very happy to be a Canadian astronaut." He may never get behind the controls of the space shuttle despite his test-piloting skills, but that doesn't bother him because he has set his cap for a space station assignment. "I'd love to fly the shuttle, but I've got lots of flying time, so I'm not proving anything by flying one more type [of vehicle]. When the space station gets up there, the really interesting work is going to be on orbit. I want to go up there for six months and work in that place." In fact, the flying machine that most intrigues him is the manned manoeuvring

unit, the thruster-powered backpack that astronauts sometimes use while working outside the spacecraft during extravehicular activity. "You're free-flying in space," he said. "That's
hard to beat."

*D*afydd *(Dave) Williams*, 37, received a bachelor's degree,
two master's degrees, and his medical degree from McGill
University in Montreal. He has received numerous professional awards and honours and is a member of half a dozen
medical organizations, including the Royal College of Physicians and Surgeons of Canada and the Canadian Association
of Emergency Physicians. At the time he was selected for the
astronaut program, he was acting director of emergency services at Sunnybrook Health Science Centre in Toronto, as well
as medical director of the advanced cardiac life support program and coordinator of postgraduate training in emergency
medicine. He is also a computer software developer. His recreational pursuits include parachuting, kayaking and canoeing,
skiing, and scuba diving. He had also started flying lessons just
before being selected as an astronaut.

Williams has been interested in the space program since his
youth, but never really thought that being an astronaut was
possible for a Canadian. And at that time, his interests were
leading him in the opposite direction. "In those days, I was
more interested in being a marine biologist and scuba diving
and living in underwater habitats." Williams started lifeguard-
ing while in his teens, and during his 20s was involved in
lifesaving activities, CPR training, and teaching emergency
rescue procedures to paramedics and firefighters. Later, he
taught CPR to flight attendants for Transport Canada and was
a member of an Ontario government committee overseeing
the use of air ambulance services. And since Sunnybrook is
the air ambulance base hospital for southwestern Ontario, he
has had a lot of direct experience with remote medicine
operations.

The lifesaving activities "sort of naturally lead into emergency medicine as a career choice, and I thought it would be
ideal to combine my basic science background in neurophys-
iology with doing research in resuscitation," Williams said. He

believes this was a major reason he's an astronaut today. "The diversity of the background that I have in emergency medicine and research was, I think, one of the factors that was important in selecting me. In terms of medical specialties, emergency medicine is probably the best. Also the ability to handle stress, the ability to deal with multiple problems simultaneously, which is something we do every day."

He hopes to continue this kind of research as an astronaut, investigating the challenges of providing health care to astronauts on long-duration space flights. This is something that all countries involved in the space station program are concerned about because the longer the missions, the more likely it is that astronauts will have accidents or become ill. In addition, there are numerous physiological changes that occur in the human body as a result of living in weightlessness in space. "I believe that now that we're looking at putting people in space for prolonged periods of time, there is a moral obligation to look at the provision of health care in space," says Williams. "The whole gamut of medical problems can occur in space; they can run the range from relatively simple right through to major life-threatening problems." He is particularly interested in the development of new medical technologies and treatment procedures for use in space.

*R*obert Stewart, 37, is an associate professor of geology and chairman of exploration geophysics at the University of Calgary, as well as director of the Consortium for Research in Elastic Wave Exploration Seismology. He earned his undergraduate degree at the University of Toronto and his Ph.D. at the Massachusetts Institute of Technology (MIT). Stewart has won many awards and honours for his scientific work, including the Chevron Oil Fellowship that supported his studies at MIT. He has lectured at universities around the world, including Cambridge University in England, and received numerous prizes and medals. He has a pilot's licence, and other recreational activities include scuba diving, hang gliding and skydiving, squash, hockey, and tennis. He is also a ski instructor with the Alberta Association for Disabled Skiers. In 1976, he was a member of the Himalayan expedition with the Alpine Club of

Canada. He is also a musician (flute and saxophone) and a singer.

Stewart, who has been interested in the space program since he was a teenager, made an unsuccessful bid to join the astronaut corps in 1983. He felt he had a better shot at it the second time around. "I got the feeling this time that it was a little different — they were looking for a broader range of candidates . . . more science, not just MDs and engineers," he said at the time his selection was announced. "And in the intervening nine years, I've done a lot more of the things they were looking for — the pilot's licence, diving, becoming more of an established scientist and communicator."

But then, after spending months jumping so many hurdles to win the prize, Stewart dropped a bombshell by turning it down. About two weeks after his selection was made public and after going through some deep soul-searching about what he wanted to do with his life, he flew to Ottawa to inform CSA officials that he'd decided to continue with his university research career. "It was the most difficult, trying professional day of my life," he commented afterwards.

He admits his decision was influenced by uncertainty about whether he would ever fly in space. Like the others, he'd hoped to be one of the Canadian astronauts sent to NASA for mission specialist training, but in late June, CSA officials decided that Marc Garneau and Chris Hadfield would be assigned to Houston. Although the decision had not been made public at the time Stewart withdrew, he knew the mission specialist option was no longer open to him, and this was one of the factors that tipped the scales. Nor could he see any clear opportunities for other shuttle or space station flights. "The summer of 1992 was a pretty uncertain time in the space effort," he said. "It's very difficult to predict what [it's] going to be like in the year 2000 . . . but it looked to me like it was pretty uncertain that I would fly before the turn of the century. Unfortunately, I had to make a very definitive decision straightaway."

For him, the central question was, "What is my contribution going to be in the prime of my career?" Joining the astronaut program entailed a 10-year commitment, and he gradually came to the realization that it would mean virtually abandoning the scientific career he loved. When he first entered the astronaut competition, he'd expected to find "a more certain area of science that I could be working in, if [I was] not flying. That uncertainty was never really resolved in my mind — exactly what I could be doing to contribute. One of the things that became apparent, that I wasn't really aware of at first, was the intensity of the astronaut program itself. It's very much a full-time job, and it would be pretty unlikely that I'd be directly able to do geophysics for a few years anyway. Because of the operational demands and because my expertise is a bit outside the program . . . it seemed to me that I would definitely have to stop my career. I didn't feel it was likely I would get back in for a long time, [so] I was not even putting on hold my other career; I was changing it period."

Counterbalancing these sobering calculations was the fact that the University of Calgary offered to extend his term as chairman of exploration geophysics and the fact that he was the leader of a research team, including numerous students, that depended on him. "In the short term, there was certainly going to be some dislocation . . . although I was trying to work out an orderly transition." On the other hand, with the astronaut program, "there were lots of good people to step in. I didn't feel like the program would collapse without me. So there was a difference in terms of where one was most needed. It came down to two great possibilities. I added up the numbers, and it weighed for staying here. I wanted to make sure that I could do creative science for the next several years. It was more a case of having hot running shoes than cold feet — I'm not an observer, I like to participate."

"He was a key person in quite a number of commitments, and . . . there was just a very powerful demand that he stay [at the university]," Aikenhead said. "On the other hand, there was a powerful demand that he come here. He tried to balance one against the other. He wanted to see what flexibility we had." CSA might have been able to delay the start of training for Stewart to give him more time to wrap up things in Calgary, but "it was clear that it was bigger than that," Aikenhead said. "We could have done a lot, but there was a limit to how much we could have done. So we agonized over it and came to the inevitable conclusion."

Asked why he waited until after being selected to make this decision, Stewart said that during the competition, he felt he couldn't get enough information about flight opportunities or what he would be doing in the program. "As a senior person with a pretty big career, and taking the 10-year commitment exceptionally seriously, I was trying to solicit a lot more information, not to seek advantage, but to understand. In fact, I'd tried to get in touch with some of the astronauts to discuss these details, but the process was so competitive that we weren't really given much access." But this was not unreasonable, he said, since there were thousands of people who wanted the job. "There's a lot of effort in recruiting, so why make more work for yourself?"

When it was suggested that there were at least 5329 people who might think he's nuts for turning down the chance of a lifetime, Stewart laughed and said, "Probably a few more than that." But speaking several weeks after the decision, he said he'd had no second thoughts and felt he had made the right decision. He joked, however, that "if the economy booms and the station is supported and there are 10 Canadian flights in the next five years — major screw-up." He said he's still an enthusiastic supporter of the Canadian space program and hopes to work with CSA on some projects. He's particularly intrigued with the potential of using the vantage point of space to study the geophysical properties of the earth. "I'm very positive about the expansion of imaging the earth from space."

He was asked if CSA officials had been irritated by his reversal after they'd publicly announced his selection and whether they'd asked why he was doing this to them. "No, I didn't get that sense at all. They said, 'Do you know what you're doing to *you*?'"

Aikenhead said that he wasn't annoyed with Stewart for making such a public about-face. "It was more disappointment, a sympathetic kind of thing — what a tough thing for him to have had to do. It took guts to do it, really, and real commitment. In the final analysis, it was loyalty to an organization of which he was a key member and to people that were depending on him."

And, although CSA officials regretted Stewart's decision, it was not as if they had a problem filling his place. All it took was one phone call.

*M*ichael McKay, 29, a captain in the Canadian Forces, was sitting at his desk at the Collège Militaire de St. Jean on Thursday, July 2, when he received a call from Bruce Aikenhead. "He told me, 'Can you talk? Can you sit down?' I was thinking, 'What's going on?' and his next words were, 'How would you like to come and join us?' I stammered out, 'What happened?' and he told me about Rob Stewart's decision. My first reaction to all this was, 'You want me to start Monday?' All I could think of was this military organization that has to get into gear to get me a posting message, and all the administration just came tumbling into my mind. [I was] stammering, 'Whoa, Monday, I can't do this.' It took me about four seconds to say, 'What am I talking about? Of course I can do it by Monday. Heck, I'll be there this afternoon if you want.'"

After receiving congratulations and further instructions from Aikenhead, he hung up the phone and let out a whoop. "There were four people in the lab and their heads all stuck up on the other side of the partition, and I said, 'I did it! I'm going to Ottawa; I'm going to be an astronaut!'"

It did take a while for the administrative wheels to start churning, so when McKay reported for duty the following week, he wasn't quite sure whom he belonged to. Because he'd gone to university on a military scholarship, he still owed the Canadian Forces about four years of his time. As he was settling into the astronaut office, he said, "I haven't got a clear contract from either CSA or the military. Those two groups are talking, but at the moment, I'm AWOL — with the blessing of my commanding officer and the Collège Militaire." When it was suggested that the military was unlikely to stand in his way, he quipped, "My mother would beat them over the head with her umbrella if they stopped me. They have been warned."

McKay admits he had been "a bit demoralized" on June 6 when he had been told he hadn't made the astronaut corps. But his natural optimism and determination soon overcame that. Having gotten so close without really having worked at it whetted his appetite, and he decided to start preparing right away for the next competition. "Within the next two weeks, I signed up for professional organizations and started to receive journals and sort of switched my courses into space science. It motivated me to see what I could do with another five or six years of preparation." But, as fate would have it, his second chance came along a *lot* sooner than that.

McKay, who had obtained several engineering degrees from the Royal Military College of Canada, was a lecturer in computer engineering and robotics at the Collège Militaire. He'd chosen these fields in part because he'd hoped someday to work in the Canadian space program. "Knowing that they were involved with robotics, and being interested in that area myself, that's what I steered my master's degree to. I wasn't thinking of the job of an astronaut." His recreational interests include gliding, weightlifting, cycling, skiing, photography, and woodworking.

Like the others, McKay had to consider the impact of his new career on his personal life. He and his girlfriend of several years, Christine Hodge, were already facing a problem that is increasingly common to young couples: careers that keep them apart. Hodge, who has a background in political science and economics, worked in Toronto, and "she's currently trying to line up work that probably will take her overseas for a few years," said McKay. "She's interested in working in Third World development. We've been living apart now for two years and travelling a lot to see each other. We would be much happier living in the same city, [but] we are very strong, very close." When he decided to apply for the astronaut program, the two were concerned about whether their new career paths would be incompatible, but "talking about it, I am pretty confident, and I think she is now, that we'll be able to get by." And, he joked, if she ends up working in a Third World country while he ends up on the space shuttle, "at least I'll get to see her every night as I fly by."

During late June and early July, the new astronauts began uprooting their former lives and planning their moves to Ottawa. It would be a temporary home — the space agency was still more than a year away from its controversial relocation to new headquarters in Montreal — so the astronauts, along with the people in charge of Canada's space station program, were camped out in an enclave of squat World War II military buildings across the road from the National Research Council. The nondescript two-storey structures were covered with peeling white paint and ensnared in a "skunkworks" of above-ground pipes that provided steam heating in winter. The only visible icon of the space age was a satellite dish sitting outside on the lawn. A CSA official acknowledged that visitors are often startled by the curiously antiquated ambience of the place.

Inside, in far-from-luxurious surroundings, the new astronauts settled into their shared cubbyholes and adorned their desks and walls with family photos and personal mementos. On Williams's wall was a going-away gift from colleagues: a picture of the earth from space with the caption "Dave's new view of the world." Spread across McKay's wall were drawings done by the young daughters of a friend, depicting several species of little green persons described as "monsters you may meet."

The first week was hectic, what with apartment hunting, dealing with the media, and grappling with the early demands of the training program; within days of arriving in Ottawa, for example, Payette took her first flying lesson. Hadfield and Garneau, meanwhile, were scrambling to get organized for their move to Houston in early August, which, for Garneau, included marrying Patricia Soame, an Ottawa woman he had been seeing for three years. As for Hadfield, he admitted that his wife had been one step ahead of him from the beginning — she had started looking for houses in Houston in February, just after he had applied to the astronaut program. He remembered thinking at the time that this was "a trifle optimistic" but said that this was her way of dealing with the pressure. "She said, 'I have to have a plan.' As it turned out, we've ended up going to Houston, so it worked out fine." He added that she had so far managed to refrain from saying, "I told you so."

Technically, the new members of the team were not astronauts yet, said training officer Parvez Kumar. "They're astronaut-candidates until they've finished the basic training program and are found fit to become astronauts." Even though they felt analyzed to death during the recruitment campaign, the intense evaluation of their skills, talents, and personal qualities would continue throughout the first year. "During the selection process you have them for a short period of time, and you're able to observe their personalities, but you really don't know if they can function well as team members, if they can function well in unusual working environments like weightlessness . . . or if they handle stress in an office environment," said Bob Thirsk, who represented Canada on an international working group that set criteria for future space station crews. After the first year, CSA would "decide who of the Canadian astronauts would be suitable space station astronauts, and at that time we'd send down one or two, or however many are required, to the Johnson Space Center, and they would become part of an international space station pool."

With the first group of astronauts, the training plan evolved on the fly, especially because the unexpected intervention of Marc Garneau's flight within the first year of operation placed special demands on the program. With the second group, things were different. Kumar had a well-organized basic program waiting for three of the new recruits when they arrived. (Chris Hadfield, who was sent to NASA almost immediately for mission specialist training, would follow a different path and would have to pick up much of what was included in the Canadian program largely on his own.) It had several objectives: to bring astronaut-candidates from widely diverse backgrounds to a common minimum level of skills and knowledge so they could be formally accepted as astronauts not only in Canada, but by the international partners as well; to qualify them to take the advanced training program leading to shuttle and space station assignments (and possibly to assignments aboard the Russian *Mir* station as well); to fulfill the needs of the Canadian astronaut program; and to maintain their existing skills.

One of the most important requirements was to give Canadians the prerequisites for NASA's advanced training programs for shuttle mission specialists and for space station operators and scientists, whose duties will be similar to those of mission specialists — operating, maintaining, and repairing systems; performing EVA; and controlling manipulator arms. (Station operators will concentrate mostly on the systems that make up the space station itself, while station scientists will be more concerned with scientific experiment packages and payloads. Another category of astronauts — payload scientists — will be roughly equivalent to the shuttle payload specialists, responsible for more specialized scientific work that needs an expert touch.)

Because Canada had built the shuttle's manipulator arm and was then in the process of building the station's manipulator, the Canadian training program placed particular emphasis on the skills needed for manipulator operation. In addition, other scientific and technical fields in which Canada has a recognized expertise were highlighted, such as upper atmospheric science, satellite communications, and space robotics. The training programs of other international partners emphasized their particular contributions — the Europeans and Japanese, for example, were both building laboratory modules — and Canadian astronauts would eventually receive training in those countries as well.

The basic training course contained a mixture of academic lectures and hands-on operational activities, including flying, scuba diving, parachuting, and participating in microgravity flights aboard NASA aircraft. The academic subjects covered topics such as space mechanics, life sciences, physics and astronomy, earth sciences, remote sensing, and materials processing in microgravity. Kumar said it's pointless to have astronauts looking through telescopes or observing the earth if they know nothing about astronomy, geology, or oceanography. He was not expecting to turn them into professional astronomers or earth scientists in one year, but the courses would give them "a background . . . to go up into space and do a good job for other people." The training program included site visits to various government, industry, and university labs where space science and engineering projects were going on. But by far the most lecture time was devoted to familiarization courses on the space shuttle, the space station, and the mobile servicing system.

A computer literacy course and an electronics workshop were also part of the program. "We'll give them computer kits or radio kits to put together," said Kumar. "It's really to get them used to the fact that they're going to be in an electronics world." In fact, all were required to meet the highest level of proficiency in the use of computers. "They've got to understand how a computer works inside out [and] be able to do their own programming," he said. They also took a photography course, because many scientific and engineering tasks on space missions require picture-taking. And they had a course on procedures for air-to-ground voice communications to prepare them for the day when they'll have to talk to mission control from the shuttle or the space station.

First-aid and CPR courses were mandatory, and fully 20 percent of the scheduled training time in the first year was earmarked for language instruction.

They also took a public-speaking and media-training course, which, interestingly, took up more time than most of the individual science lectures. Canadian astronauts generally receive more exposure to the media than their U.S. counterparts. "NASA doesn't allow new astronauts in front of the press for at least 40 weeks, almost to the end of the first year's training," Kumar said. "I feel that's a little too restrictive." On the other hand, they have to protect their time. "I've said that anything that conflicts with the training program is out. By and large, the PR is to be kept to a minimum."

Maintaining physical fitness was a priority item in the training program, since the astronauts must continue to meet NASA's mission specialist standards. Consequently, about 16 percent of their scheduled work hours was set aside for this. And then, as if their plates were not heaped high enough, there were flying, scuba diving, and parachuting lessons. The inclusion of these activities was a new departure for the Canadian astronaut program. Although three of the original astronauts (Ken Money, Bjarni Tryggvason, and Roberta Bondar) were pilots, flying was not part of their formal training and was not even particularly encouraged, although others got their licences on their own time. Later, they managed to persuade CSA officials to allow skydiving as well, arguing that since their job requires them to spend quite a bit of time flying in high-performance aircraft, they should know what to do if they had to bail out. Also, after the *Challenger* accident, NASA had installed a new emergency escape system in the shuttle that requires the use of parachutes, "and we really should know how to do this," said Money, one of the most enthusiastic advocates of skydiving. Finally, there was the argument that jumping provides good stress-management training. (At that time, Soviet cosmonauts were required to do 100 free-fall jumps before going into space.) "You discover something about yourself in the moment you actually jump and fall out of the sky," says Garneau. "It teaches you about your ability to work in potentially stressful situations and to react quickly and correctly."

At the time, CSA officials seemed to regard many of these activities as more of an indulgence than anything else, but they gave the go-ahead. Aikenhead said they'd had "some reservations" about the parachute jumping — the kind of res-ervations felt by "a parent [who] learns that the kids have decided to take up a dangerous activity." But the logic of emergency preparedness and stress-management training was persuasive, and the training program for the second gen-eration of astronauts placed considerable emphasis on these kinds of skills, in large part because they will have a far greater "hands-on" role aboard the space station than payload spe-cialists did on the shuttle. Kumar said he pushed hard to have flying included in the curriculum precisely because of the operational skills it requires, such as good eye-hand coordina-tion and the ability to handle complex machinery in demand-ing circumstances. Dave Williams, Julie Payette, and Mike McKay were therefore expected to acquire at least a private pilot's licence and had the option of going even further in their flight training. (As the U.S. Navy's test pilot of the year, Chris Hadfield was naturally exempt from having to do endless touch-and-go landings in a Cessna.) They also had to learn to fly a glider up to the solo level and reach the free-fall level in parachute jumping.

Scuba diving skills were directly relevant; one emergency evacuation exercise, for example, involves being dropped into the water and scrambling into a life raft while wearing a heavy evacuation suit. Bondar got her scuba certificate while in training for her mission, saying it would help her reduce the unfamiliarities of having to pull off such a procedure for real. Scuba skills are also useful to astronauts training for space walks because they work in a large water tank (known, appropriately, as the WETF or Weightless Environment Training Facility) that NASA uses to simulate some aspects of the slow-motion movement of arms and legs that characterize working in a bulky space suit in weightlessness.

In total, the training program encompassed 1875 hours (50 weeks), of which more than 1600 hours (43 weeks) were earmarked for scheduled activities. The other seven weeks were unscheduled for flexibility. Although the program was designed for the new astronauts, members of the original group could also take courses they were interested in, Kumar said. "They've looked at it and said, 'If only we'd had this.'" He added that both NASA and the European Space Agency have shown interest in imitating some parts of the program.

After the first year, the astronauts would spend another year of advanced training at facilities run by NASA and the other international partners. "I want to make sure, because of the cultural aspects of space station, that all our astronauts spend some time in different countries," Kumar said. "It's not so much of a problem with the States, but certainly with the Japanese and the Europeans, we would give them a briefing on the country, the people, the customs. But ... they are highly intelligent, and with their motivation, they will pick it up. If they don't, it will be a message to us saying maybe this person is not suitable to go on the station, but will be okay on the space shuttle."

During the second year, the astronauts would also be allowed to spend time on their own projects, such as practising medicine or doing research in their own fields of expertise. Even though the original astronauts tried to stay active as scientists and engineers, the demands of their workload meant they inevitably lost a lot of the skills they'd brought into the program. Kumar said he didn't want that to happen to the second group. "As long as they're occupied, they're in touch with the user community, or actively involved in some of their own work. If they can produce papers, so much the better."

He added that even though the new astronauts were hired for space station assignments, it's important that they qualify to fly shuttle missions, too. "My feeling is that NASA will not allow rookies on the station the first time around," said Kumar. "In order for us to have our allocation, we need to make sure that all our astronauts have had at least one shuttle flight under their belt, or preferably two. That's why I think we have to concentrate on the shuttle."

Speaking during the first week the new group was in Ottawa, he commented that they were "still on a high and I expect them to be on a high. You have to allow them that luxury." But not for long. He promised that he'd soon be telling them, "Now, down to reality."

Amid the excitement of embarking on their new lives, the new astronauts paused to assimilate the startling news that Rob Stewart had turned down the adventure on which they were embarking. They'd had no inkling it was coming, and none of them had spoken to Stewart about it. Payette wished they'd had the chance. "If Rob had talked to us, maybe we could have brought in the positive perspective. It is a big commitment, yes, and maybe he had the most to lay aside. But Dave also had a lot to lay aside in order to accept this job. Maybe [Rob] made his decision too soon. If he had come here, he would see that our flight opportunities today are better than they have ever been for Canada."

Hadfield also expressed the wish that "we could have been with Rob during his decision-making process. He made his decision in seclusion from this whole influence; he was back in his other environment, looking around at his office and his friends and his future and everything else. But obviously he made what he thinks is the best life decision for himself. And, of course, we are happy to have Mike on board."

While the new astronauts were having a hard time coming to terms with Stewart's decision, it was less inexplicable to some of the veterans, who'd experienced unexpected years of waiting. "When we came in, it was a three-year program; that was something you could afford to take," said Bjarni Tryggvason. "Now that it's a 10-year program, people have to be much more concerned about what they're going to be able to do in the program, compared to what they were doing outside."

As a member of the selection committee, Marc Garneau was disappointed by Stewart's decision. "I thought he had already made that adjustment in his mind . . . that he had sorted out how he was going to relieve himself of his previous commitments." But he too acknowledged that "it's a big adjustment and a risky one. As much as we were trying very hard to select the right number of people so that each one would have a reasonably high expectation of flying, there was no way we could guarantee these people a flight. So they were potentially giving up a lot to potentially gain a great deal. It really does require a lot of soul-searching."

Roberta Bondar expressed similar sentiments. When the news broke, she was quoted in the newspapers as saying, "Sure, a space flight is a great thing, but you have to weigh these things for yourself. When you are established in a certain profession, there's an awful lot for you to give up." She knew what she was talking about, because she'd had eight years in the prime of her research career during which to ponder what she'd given up as a scientist to realize her dream of flying in space.

But, in the end, she was one of the fortunate ones: she got her flight. Most of the others had not, nor did they know when — or if — they ever would. For them, the countdown clock was still on hold.

2

Waiting

People are asking, "Well, what are you doing?" It's in their minds that unless you are flying, you're sitting around twiddling your thumbs.

— MARC GARNEAU

More than seven years separated the shuttle flights of Marc Garneau in October 1984 and Roberta Bondar in January 1992. The Garneau flight had been an unexpected bonus; the astronauts had originally been hired to fly just two shuttle missions, then scheduled for 1985 and early 1986. As it turned out, Garneau's flight was all that kept them from spending almost a decade without getting anywhere near going into space. During the early 1980s, NASA's increasing difficulty in keeping the trouble-plagued shuttle program on track kept pushing the two original Canadian missions further into the future, and then in January 1986, *Challenger* exploded in the brilliant blue sky over the coast of Florida and all bets were off.

Years later, as she was on the verge of becoming only the second Canadian to fly in space, Roberta Bondar commented, "I think in retrospect that Marc was exceptionally lucky that he came in at the time he did and flew so quickly. That's a phenomenon never to be repeated for years, I'm sure."

Because the Canadian astronaut program was so closely tied to the fortunes of the shuttle, the months after *Challenger* were a period of great uncertainty and confusion. It soon became distressingly obvious that fixing the shuttle's technical problems would take time. Initial optimism about an early resumption of a regular flight schedule faded, and like everyone else who depended on the shuttle, the Canadian astronauts suddenly found themselves facing a delay that was likely to be measured in years rather than months. They had little option but to wait it out. Even though there was a very active Soviet manned program at the time, and political events were making it increasingly accessible to Westerners, there were no mechanisms in place to allow Canadians to jump on board in a hurry, even if funds could be found. Garneau's flight and preparations for MacLean's mission, which had been scheduled for early 1987, had pushed the resources of the astronaut office to the limit, and they'd had time for little else. Suddenly, after *Challenger*, all this furious activity was put on hold, and no one, including the astronauts themselves, was entirely sure what they should or could be doing instead.

The early part of 1986 was unsettling for other reasons as well. At that point, the Canadian astronaut program still existed officially for no other reason than to fly the two shuttle missions. One mission was designated an "engineering" flight to test a new machine vision system to improve the performance of space manipulators. The other was a "life sciences" flight focussing on experiments designed to increase scientific understanding of the effect of microgravity on human physiology. (Since there are very tiny gravitational effects in space, scientists typically use the term "microgravity" rather than "zero gravity.")

Although from the outset, there had been the hope and intention of creating a permanent astronaut corps, the long-term fate of the program was closely tied to Canada's involvement in the space station program. At that time, however, Canada was still embroiled in difficult negotiations with NASA concerning its contribution to the international project. The U.S. space agency, seeking international partners partly to cushion anticipated financial problems at home, had invited Canada, Japan, and Europe to join in building and operating the station. The decision was a major one for all three potential partners; not only were big bucks involved, but a commitment to the space station program would require wholesale reorientation of their national space programs.

Canada proposed to build the station's robotic manipulator system, known as the mobile servicing system (MSS), a project that would not only capitalize on its experience in developing the shuttle's manipulator arm, but also would promote additional research in Canada in important new high-tech fields such as automation, robotics, and artificial intelligence. An agreement on Canada's space station participation was finally struck in mid-March 1986, but political complications, stemming from opposition to the Canadian project by the U.S. robotics industry and its political supporters, continued to plague negotiations between the two countries throughout most of that year.

For the astronauts, the most significant aspect of the space station pact was that, in return for its contribution of the MSS (which was to cost Canada an estimated $1.2-billion through the year 2000), Canada was entitled to three percent of the crew time aboard the station, amounting to approximately six months every two years. (This calculation was based on a crew of eight astronauts; however, the financial problems faced by the space station program in the late 1980s forced NASA to reduce the planned crew complement to four during the early years, so Canada's share will be reduced proportionately. "Everybody took a hit," said Bruce Aikenhead.)

There was also an understanding that there would be at least one more shuttle mission for a Canadian astronaut, on a flight associated with checking out the operation of the MSS before it begins the task of helping to assemble the space station. The agreement ensured a future for the Canadian astronauts, and by mid-May, the government announced that they would become an integral and continuing part of the Canadian space program.

But even though the program had been put on a more secure footing, the post-*Challenger* period forced the astronauts to face some difficult personal questions: How long would the delay be? What would they do in the meantime? What were their chances for shuttle flights or space station missions afterwards? What would all this do to their scientific careers? In short, how much of their professional and personal lives were they prepared to tie up in the astronaut program, given the uncertainties that now confronted them as to when — or whether — they would ever fly in space? Adding to the uncertainty, the space station program itself kept running into financial and political problems in the United States that threatened to stretch its construction schedule beyond the turn of the century. In fact, the difficulties were so severe that at times it seemed outright cancellation could not be ruled out — a problem that persists into the 1990s.

Each of the six Canadian astronauts had to weigh factors such as age, career ambitions, and the probability of getting a flight, along with an even more personal re-evaluation of the risks involved in flying in space and their potential effect on families and friends.

Steve MacLean had the advantage of already having been selected for one of the two scheduled shuttle flights. As one of the three "engineering" astronauts in the group, he would be testing the new vision system. At 31 the youngest of the astronauts, he could also reasonably expect a shot at subsequent shuttle or space station missions. The same was true for 32-year-old Bob Thirsk, who had been the backup astronaut for Garneau's mission, a tantalizing but frustrating experience that involved spending an intense year preparing for a shuttle mission he was never likely to fly. Although the training had been exciting in its own right, it fell short of the ultimate payoff, and this made him more determined than ever to have a flight of his own. Thirsk, along with Roberta Bondar and Ken Money, made up the "life sciences" contingent of the astronaut corps, and they were in the running for the other shuttle flight, which had yet to be assigned at the time of the *Challenger* accident.

For Bondar, then 40, the post-*Challenger* period was a time of soul-searching. Her father had died suddenly shortly before the accident, an event that hit her especially hard because he'd always been a champion of her efforts, especially her interest in science and math, from the time she was in school. It caused her to ponder anew how the risks of her job might affect her family, to whom she is very close. "After the accident, I had to regroup my thoughts," she said. "It was a time of sorting through and thinking. It's one thing to have your own goals, but it's another thing to be putting other people at an emotional disadvantage in case something does happen."

For Money, the oldest of the astronauts at 51, the delay dealt another blow to his dream of decades. He'd been doing research in space physiology for nearly 30 years and had long harboured the hope of getting a space flight. In fact, long before there'd been a Canadian astronaut program, he'd come tantalizingly close to being selected for a NASA flight, only to lose the opportunity because he was a Canadian. Since Money did not plan to be in the program when space station assignments were being handed out some 10 to 15 years hence, he regarded the shuttle life sciences missions as his last shot at flying in space.

Bjarni Tryggvason, 40, was in training as the backup astronaut for Steve MacLean's flight and was also involved in developing equipment for that flight. Like Thirsk, he was intensely involved in preparing for the mission but was not expecting to fly it. Similarly Marc Garneau, then 36, had no immediate flight prospects, but his one shuttle flight had made him eager to repeat the experience, so he was not seriously tempted to retire. He had set his sights on the shuttle flight on which the MSS would be launched, which was then scheduled for the mid-1990s.

Thus, each of the six found reasons to stay in the program, at least until shuttle flights resumed and the status of future missions became clearer. Reflecting on that post-1986 period several years later, Garneau commented: "You'll notice that none of us quit. Everybody had a good think about it. But if you look at the NASA astronauts, no one who had not flown quit. Some left after [*Challenger*], but they'd already had one or two flights. Those who hadn't flown, no matter how awful the tragedy was, were not sufficiently dissuaded by that tragedy to decide this is not worth it." The same was true for the Canadian astronauts. But that, he said, left them facing the question, "How do we occupy our time in a useful manner?"

It was not that they felt they had to cast around for a reason to exist at all. "There were things to do and things to prepare for," said Thirsk. "We knew after the *Challenger* blew up that our program was going to be delayed, but the two missions were still going to take place and our program was still going to be a long-term program." However, since the preparations for the shuttle missions progressed at a much slower pace, the astronauts found they had time for other things, such as resuming previous research projects or developing new ones, and expanding their role in providing support to Canadian space scientists, who benefitted from their expertise in developing and testing equipment and conducting experiments in microgravity.

These activities transformed the Canadian astronaut program. Instead of being a small, tightly knit operation devoted almost exclusively to intensive training for shuttle missions, it became a more research-oriented program, handling a wider range of tasks and interacting with a wider scientific community. Inevitably, this diversity of activities tended to take the astronauts off in different directions. "Right after the accident, everyone started thinking about their careers, and that's one reason why we sort of dispersed," said Thirsk. "We were a very tight group of people up until 1986 because we all worked out of the same office, we all did the same things. After 1986, some people went off to Toronto to pursue interests they had there, some remained in Ottawa, some got involved in space station. We're not working as a group of six on one project; instead we're six people working on 18 different projects."

One of the things Thirsk decided to do was to go back to being a doctor. He joined an Ottawa clinic as a part-time general practitioner and started seeing patients two half-days a week. Most of his patients knew about his double life. "Once in a while someone will say they saw me on TV or they read an article about me, and we end up spending the first five minutes talking about space," he said. "I don't mind because it's an important thing for a doctor to develop a good relationship with a patient, and I find that someone who doesn't know one end of the shuttle from another is still interested in the idea of exploring space. So it's a good way to break down some of the barriers between a new patient and a doctor." On the other hand, he had to explain that he might not be around as much as the other doctors. "Sometimes I have to cancel clinics at the last minute because I have to go off here or there. It might be nice to be able to say your doctor is an astronaut, but the disadvantage is that I am not quite as available as someone else."

Thirsk valued the clinical experience for several reasons. He'd originally gone into medicine, after obtaining two engineering degrees, with the intention of pursuing a research career in biomedical engineering. "But once I got into medical school, I became more enthused about the patient contact aspect of medicine. By the time I finished, I was hoping to remain with my original objectives, but I didn't want to give up the clinical side." He says his part-time medical practice provided the human contact that helps him remember that, as a researcher, "your objective has to be to meet the needs out in the world, the social needs." He also believes continuing clinical experience is important to his career as an astronaut because NASA plans to have at least one medically trained astronaut on space station missions. "So I am maintaining my skills. If I don't stay in clinical practice, I would quickly lose [them]. I think it's a very important thing to hold on to, and I regard it as part of my astronaut training."

At the same time, he began work on a research project that combined his interests in medicine and engineering — designing a new version of the anti-gravity suit astronauts wear when returning to earth. The suits are worn to protect against faintness or blacking out on returning to earth, as gravity pulls blood to the lower part of the body — a danger that increases after spending time in space because the cardiovascular effects of microgravity include a reduction in the total volume of blood and other fluids in the body.

Thirsk's development of the new suit was an offshoot of research he'd done while working on a master's degree in engineering at the Massachusetts Institute of Technology (MIT). There he had been part of a team trying to develop a noninvasive method of preventing blood clots in the leg, a condition known as deep-vein thrombosis that can cause problems for postoperative and bedridden patients. As an alternative to the traditional method of using anticoagulant drugs, the MIT researchers were trying to develop a mechanical method of pushing blood out of the veins in the leg to the upper part of the body. "One of the criteria to prevent thrombosis or clots from forming is to empty the veins periodically," said Thirsk, "so we were working on a mechanical device that would collapse the veins and move the blood out of the legs. When I entered the astronaut program, we had a similar problem — not the clots forming in the legs, but the need to move blood out of the lower extremities up to the central circulation. I was able to use some of the techniques we had developed for the problem of deep-vein thrombosis and incorporate that into an experimental antigravity suit, to try to get the some of the venous blood flowing back up into the central circulation region where it's really needed." Testing the prototype of the suit, and associated measurements of blood flow and volume in the veins of the astronauts' legs, were among the experiments done on Bondar's flight in January 1992.

Among his other duties, Thirsk also served as Canada's astronaut representative on two international working groups, one studying the selection criteria and training requirements for space station astronauts, and the other concerned with space station medical standards. He has also been part of an international working group studying life sciences issues associated with future manned missions to the moon and Mars.

In the first year after the *Challenger* accident, Roberta Bondar spent much of her time developing contacts for the Canadian astronaut program with space researchers in Japan and Europe. Concerned about the difficulties of obtaining funding for expensive new space projects, she believed that collaborating with scientists from other countries could help Canadian researchers who wanted to develop space experiments. "I felt there was a need to develop international opportunities. But after a year of getting things organized with all these people, I decided it was high time that I looked at my own research. Enough of this being a scientific philanthropist."

Like Thirsk, Bondar wanted to keep her hand in as a doctor. She started spending time in Toronto, seeing patients at

Sunnybrook Health Science Centre and Toronto General Hospital. But she is also strongly research-oriented, and she decided to pick up on the stroke research she'd been interested in prior to becoming an astronaut. In fact, the project she settled on dovetailed nicely with her work in the space program. The link was provided by a new method of studying blood flow in the arteries of the brain, known as trans-cranial doppler. This noninvasive technique, which is useful in studying patients who have suffered strokes on earth, can also be used to study changes in blood flow in the brain resulting from the shifting of body fluids that occurs in microgravity. "This is all new science," Bondar said. "For a long time, people have wondered whether or not there is raised pressure inside the head because of the shift of fluid — whether or not the brain is somehow affected."

Bondar began a research program with colleagues at Ryerson Polytechnical Institute, Toronto General Hospital, and the Johnson Space Center. One part of the study involved running tests on subjects aboard the KC-135 (NASA's microgravity training aircraft), and Bondar says the results suggest that, in microgravity, there are changes in the output of blood from the heart that alter the blood-flow pattern in the brain. Much more research is needed to understand what this means, but Bondar hopes these studies will ultimately provide a better understanding of what happens to the human body in space. They should also benefit earth-based medicine by "helping us understand a bit more about some of the basic physiological things that are going on normally in you and me just walking around."

It was not a difficult transition for Ken Money to resume his research in the post-*Challenger* period because his work had always been closely tied to the space program. He is a well-known expert on space motion sickness, a malady caused by the effect of microgravity on the brain and the balancing system in the inner ear. Surprisingly, it afflicts to some extent at least 70 percent of the astronauts who go into space. Actually, Money says, while 70 percent admit to experiencing motion sickness, the true number is probably closer to 90 percent. About 30 percent experience vomiting, while another

60 percent have less severe symptoms. "Everybody minimizes motion sickness when reporting it," he said. The symptoms usually last two or three days, although some astronauts have been affected throughout a shuttle flight and one Russian cosmonaut was afflicted for 14 days. On shuttle flights, which typically last only seven to ten days, the loss of productivity can have a serious impact on the scientific work, and there's also concern about the safety implications if an astronaut were to be sick during critical operations, such as EVA or returning the shuttle to earth.

One of the puzzling aspects of space motion sickness is that it occurs in people who do not normally experience motion sickness on earth; in fact, many of the afflicted astronauts are skilled pilots who can put high-performance jets through their paces without a twinge. This makes it difficult to predict susceptibility to the problem, or who will actually get sick in space. For nearly 30 years, Money had been working to understand the disorder and to develop predictive tests and new treatments for it. He had conducted many experiments with NASA crews before becoming an astronaut himself and also developed motion sickness experiments conducted by Marc Garneau on his shuttle flight.

During the last half of the 1980s, while waiting out the post-*Challenger* delay, Money spent most of his time in the lab at the Defence and Civil Institute of Environmental Medicine (DCIEM) in Toronto, where he'd worked for many years. (Technically, he was still on staff there, on loan to the astronaut program.) The next five years were scientifically productive — he was involved in numerous studies with his DCIEM colleague Bob Cheung and authored or co-authored more than two dozen scientific papers. "I eventually published more papers than I used to before. I had better access to travel funds and better access to the KC-135." In one study, Money and his colleagues found that people who did no regular exercise were less susceptible to motion sickness. "If you're a couch potato, you're resistant to motion sickness." Asked what explained this finding, he laughed and said, "We really have no idea."

As Canada's representative on the International Academy of Astronautics, Money also participated in a study of a manned mission to Mars, examining the effect that such a trip would have on the physiological and psychological well-being of the crew members.

Steve MacLean, Bjarni Tryggvason, and Marc Garneau continued preparing for MacLean's shuttle flight, the primary goal of which was to test the prototype of a space vision system (SVS) that would improve the speed, accuracy, and safety with which astronauts could grab objects with the Canadarm (see Chapter Five). This work included getting the system through numerous tests in NASA simulators, all of which went well. "If we had failed down there, NASA would have been a little bit nervous about flying our system," said MacLean. "I think we impressed the hell out of them."

On his flight, MacLean would be responsible for six other experiments as well, so he also spent time working with the scientific investigators who designed the experiments, learning to use the equipment. He tested one device, designed to take measurements of chemical processes that affect the earth's ozone layer, by taking measurements from the cockpit of a commercial airliner as it flew over the North Pole. The flight took about 10 to 12 hours, and "it was constant data-taking; you didn't stop for the entire time," said MacLean. These experiments actually benefitted from the shuttle flight delay, he added. "Every experiment has improved tremendously because of knowledge we've gained in the interim. Many people don't appreciate what you're able to accomplish even though you don't fly in space."

The delay in his shuttle mission also gave MacLean an opportunity to do something dear to his heart — continue research on a second-generation SVS. He became manager of a program to develop an advanced SVS incorporating new laser technology and three-dimensional video imaging, which he hoped would one day be used not only with the shuttle's manipulator, but also with the mobile servicing system on the space station. A laser physicist before he became an astronaut, MacLean was particularly intrigued with the idea of developing a sophisticated laser ranging-and-scanning device to improve the system's capabilities.

As the backup astronaut for MacLean's flight, Bjarni Tryggvason was also involved in all the training activities. In addition, he was responsible for the design of the Canadian target assembly (CTA), a metal plate with special markings that would be manoeuvred and grabbed by the Canadarm during the SVS tests. His other activities included teaching university engineering courses and serving on committees studying advanced technologies for the space station and the development of training facilities for astronauts who will operate the station's manipulator system.

As the time for MacLean's flight got closer, Tryggvason assumed responsibility for shepherding the CTA through the long and arduous approval process NASA requires for anything that goes into the shuttle cargo bay. He said that Garneau's mission had been a useful learning experience "because it gave us a very quick and thorough look at how NASA works and all the things you have to do to get equipment on the shuttle, all the safety-related questions you have to answer. [But] we're putting a lot more equipment in this time; we're putting stuff in the cargo bay, which we had not done before, and so there are new things we're learning about what's involved with that."

Garneau also got some first hand experience with NASA's paperwork jungle when he took over the responsibility for getting SVS equipment approved for use in the shuttle crew compartment. To ensure the safety of the crew, equipment used anywhere inside the shuttle must meet stringent standards, but the requirements for the SVS components were even more onerous because they were being installed on the flight deck amidst the shuttle's control and navigation systems. "You have to satisfy requirements with respect to toxicity, flammability, no sharp edges and corners, structural integrity, no electromagnetic interference," Garneau said. "You could write a book just to clear a pencil for flight. The system thrives on paper." He likened NASA's gruelling safety reviews to a court of inquisition. "I felt like Joan of Arc, going in front of this group of about 25 people whose only mission in life seems to be to find a flaw in your argument or to shut you down. The first time I did it, I would say there was blood on the floor. A lot of it was mine, but some of it was theirs." However, he came away from that first encounter "feeling that I know what these people want. And we had another one, which went very much better."

Asked if this was how he'd envisioned an astronaut's life when he first joined the program, Garneau chuckled. "Well, I had a feeling after my flight, which I enjoyed so much, that I was going to have to repay in some way. And so every time I say to myself, 'Boy, oh, boy, is it all worth it?' I realize that, yes, of course, it is worth it. It's not just for Steve's personal satisfaction of going up into space — it's really because it is important to Canadian scientists that his experiments work up there and that we don't endanger anybody's life in the process."

The years between flights also brought changes in the personal and family lives of the astronauts. The three who'd been bachelors at the outset married and started families. Bob Thirsk and his wife, Brenda Biasutti, who'd married shortly after Thirsk joined the program, had a daughter in 1987 and a son in 1990. Bjarni and his wife, Lily-Anna Zmijak, had a son in 1985 and a daughter in 1988. Steve MacLean and Nadine Wielgopolski had a son in 1990 and a daughter in March 1991, around the time MacLean had to move to Houston to begin final training for his flight. Because Wielgopolski was on maternity leave, she and the children were able to move with him during the last six months.

Marc Garneau's family life took a tragic turn in 1987, when his wife, Jacqueline, from whom he was separated at the time, took her own life. In the years that followed, Garneau faced all the complexities of being a single father of teenage twins while juggling a demanding job that required a great deal of travel. "I try to plan my travel so that I can do the most work possible in one trip rather than spreading it across two or three trips," he said. "I feel I owe it to my family to plan it in the most efficient possible way. The result is that I don't think I'm away any more than a large number of civil servants who are called upon to travel." Garneau remarried just before moving to Houston to begin mission specialist training in the summer of 1992.

Perhaps the most unusual media controversy concerning an astronaut's private life focussed on children who didn't even exist. As the years wore on, Roberta Bondar remained not only the sole female astronaut, but the only unmarried and childless one. As with other demanding professions, being an astronaut forces more difficult choices on women than it does on men when it comes to balancing careers and personal lives, and Bondar was asked the inevitable questions about whether she felt she'd had to sacrifice having a family to stay in the astronaut business. She acknowledged that, to some extent, the questions reflected a legitimate debate about the juggling act required of women professionals. She also found they verged on invading her privacy. Nevertheless, she gamely tackled the question whenever it was thrown her way. On one occasion, she reflected that she probably could not have combined motherhood with her work in the astronaut program, particularly since she would have wanted to spend some time at home with her children. Certainly, she said, the demands of waiting and preparing for a much-delayed shuttle mission had precluded anything resembling a normal social life.

Stephen Strauss, science writer for the *Globe and Mail,* took up this issue with some vigour. In an editorial headlined "Roberta Bondar gave up much to be an earthbound astronaut," he discussed why she was "forced to make a personal decision unlike any of her nonflying male compatriots." He quotes Bondar as saying that when she'd entered the program at 38, she'd been "naive" enough to think she could "spend two or three years in the program and still have time to have children of my own." But by 1991, at the age of 45, she conceded that option had been abandoned, that children would not be part of her "biological contribution" to society.

Being a Canadian astronaut circumscribed Bondar's choices even more than might otherwise be the case; women astronauts in the U.S. program have taken maternity leave and returned to flight status, but they have much more frequent opportunities for shuttle missions than a Canadian astronaut. At a media briefing before Bondar's mission, Strauss asked CSA officials (all male) whether they'd given any thought to adapting the Canadian program in the future to "allow women to come in and out and have children." Clearly startled by the question, they greeted it with a brief, awkward silence before Larkin Kerwin, then CSA president, gathered himself to comment, rather dismissively, that even during the space station era, it is "not anticipated that astronauts will be spending years in space — it will still be a case of weeks and sometimes months — so that should not interfere unduly with the personal lives of the astronauts."

It was an answer that fell into the "just don't get it" category. The fact is that it takes many years to acquire the educational and professional qualifications to become an astronaut and further years of training to earn the right to spend those few weeks or months in space. Since these usually coincide with a woman's prime child-bearing years, most women astronauts, like other professionals, are forced to either delay or forego having children, or face a major interruption at the height of their career. For many women, it is a wrenching juggling act — one that men don't have to face and often seem unable to comprehend.

In one way, the hiatus in shuttle flights benefitted Canada's space research community, because it enabled the astronauts to devote much more time to providing advice and assistance to Canadian scientists who wanted to develop experiments to be carried out aboard the shuttle or space station. "The philosophy in our office is that when [astronauts] are not specifically training for a mission, we should be assisting or supporting in some manner or another," Garneau said. Thanks to the experience gained during his mission, the astronauts were in a position to offer an insider's knowledge of how the NASA system worked. For example, they'd learned all about the complexities of getting equipment approved for space flight, and they'd become well versed in the challenges of doing experiments in microgravity.

The biggest problem was that after *Challenger* there was almost no access to space and, therefore, to a microgravity environment in which to do space experiments. So the astronauts began pushing for the establishment of a Canadian R&D program using the KC-135, NASA's microgravity training plane, as a research facility. By flying a roller-coaster pattern of climbs and dives, the KC-135 produces microgravity for about 20 to 30 seconds as it goes "over the top." NASA uses the aircraft to train astronauts (who refer to it, less than affectionately, as "the Vomit Comet" because of the tendency of the roller-coaster ride to cause motion sickness) and to test equipment destined for the shuttle.

Ken Money was a veteran of the KC-135, having spent years conducting motion sickness experiments on it. "Ken really got the ball rolling [and] pushed the concept of using the KC-135," said Garneau. "This was after *Challenger*. We were in the doldrums; activities seemed to have ceased, and it looked like they were going to be slowed down for a very long time." Before long, CSA was booking KC-135 flights two or three times a year. "The level of activity within Canada has really mushroomed from coast to coast," said Garneau. By the early 1990s, the program had become "a well-oiled machine," and CSA had a waiting list of scientists who wanted to get on the flights.

One of the main advantages of the KC-135 was that it offered researchers a comparatively inexpensive way to make mistakes while learning how to design experiments and equipment that will work in microgravity in space. "We always mention to our KC-135 investigators who are flying for the first time not to be disappointed or frustrated if after the first flight their experiment is a total failure," said Thirsk. "You really have to switch your mind into a zero-g mode. You have to experience it for yourself, and that helps you improve your design the next time. Things float in space. That means not only the operator floats, but your equipment will float if it's not fastened down properly. It also has implications for the way your equipment operates; for example, you don't really know if your floppy drive is going to function as it did in a one-g environment. There may be certain components within your computer that rely upon the one-g force field to make contact between two surfaces. In zero-g, of course, that doesn't happen."

And then there are the microgravity effects on human operators that must be taken into account. For example, it's difficult for a person to stay put long enough to get things done. "Your legs are just about useless in zero-g," Thirsk said.

Hands normally used to operate equipment are needed to keep "a firm grip on something in order not to float about." This sort of thing affects equipment design, he said. "It's not only that hardware doesn't necessarily work, but perhaps the procedures you've developed in a one-g setting with a one-g perspective in your mind are just not realistic in zero-g. Maybe it takes a longer period of time to actually do the experiment or maybe it takes more than your allotted one or two people to do the experiment. So in our KC-135 program, the primary focus is to allow researchers to do some microgravity experimentation, but a secondary purpose is to allow people to check out their equipment and the procedures they have planned for a space shuttle or space station flight."

Because of their extensive experience with microgravity, the astronauts were in a position to help scientists, both before and during the KC-135 flights. "It would be just about impossible for them to design a piece of hardware for the KC-135 that would work as they think it will the first time around," said Tryggvason. "They just don't have the effects of [microgravity] ingrained into their thinking pattern. Simple things such as putting a pencil holder in an appropriate spot. And there are many more challenging problems than that."

Garneau said that "every time there is a new piece of hardware, we'll go out to the university or wherever the hardware is being developed and make sure that it's going to pass the safety review, which is pretty strict, and that they're doing it in a way that's practical — that it can actually achieve the objective they want to achieve on board the space vehicle. And we also make sure that they have no illusions about what the [KC-135] environment is like." On the KC-135 (and even on the shuttle), tiny gravitational effects occur, and while they are subjectively indistinguishable to human beings, they can affect some extremely sensitive experiments. And the KC-135 differs from the shuttle in that microgravity lasts less than 30 seconds. For some experiments, "that is a limiting factor," Garneau said. "On the other hand, there is a great deal of science that can be done and equipment that can be tested in [that time]."

The astronauts were also on board the plane to assist with

the actual performance of the experiment, especially if the researchers had never flown before. More than a few first-time passengers on the KC-135 have experienced nausea and vomiting and sometimes are too debilitated to get the job done. "We are there to supply the support in case somebody gets sick," said Garneau. "Sometimes they say, 'We don't need any help; we can do it ourselves,' but they haven't experienced the environment. Once we're sure they're going to be fine, on the second flight they can do the experiments themselves. But the first time, we insist that at least one astronaut be there."

Astronaut training officer Parvez Kumar, who was formerly head of a program that funded Canadian scientists to develop experiments to fly on the KC-135 and the shuttle, said that it tremendously speeded up the evolution of the scientific equipment when investigators were able to do their own experiments on the KC-135, but the collaboration of the astronauts played a major role in this evolution as well. "The main thrust was to get our astronauts down to the working level, really put them in place of the scientist that couldn't fly, to be able to say in flight, 'Now we know how to modify this thing, we'll make it work better next time.'" In addition, they have to be scientists themselves [and] have a thorough knowledge of what the experiment was about, not just hit a button, get a recording, and give it back to the [investigator]. It is that second level of input that speeds things. I could see that the upgrade of an experiment, when it flew again, had their input. In addition, because they are seeing a lot more of what's going on, [they can] put new ideas into the minds of a particular scientist. So they are a resource to the scientific community. That cross-fertilization is really important."

But the astronauts weren't just working on other people's problems — they developed a number of their own experiments. Tryggvason, for example, found the KC-135 program an unexpected opportunity to do some creative engineering. "I anticipated that in the first two or three years [of the astronaut program], I would not have much time for doing the kind of technical things I like to do. We were just going flat out to get ready for these flights; we had our plates full. But I was hoping that, as the program developed, I would be able to get into doing some microgravity science. The *Challenger* accident really opened that door quite wide."

Working with university students, he developed several instruments and devices to improve the ability of researchers to perform and monitor microgravity experiments. These included an infra-red system somewhat like a TV remote control to exchange data with the experiment package, and a cage-like "vibration isolation" device to prevent the package from spinning or bumping into walls while it's free-floating in microgravity. NASA was sufficiently intrigued by the vibration isolation system that it picked up the flight costs for testing it, and several U.S. companies have expressed interest in marketing the device commercially. These companies had been trying to develop similar technology, but the Canadian researchers beat them to it.

For Tryggvason, the exercise was not only a personal challenge in engineering design, but an opportunity to educate Canadian scientists and engineers about doing space research, particularly those just starting their careers. "One way of getting Canadian scientists ready for using space station is to get them when they're young, so I've done a fair bit of speaking at universities, technical presentations on what microgravity is all about," he said. "A lot of engineering and science schools have requirements in their third and fourth years that the students work on projects." Several students from the University of Ottawa, the University of British Columbia, and Simon Fraser University got project reports out of their work on his KC-135 devices, and some even got to fly on the aircraft. Four Simon Fraser students developed a working prototype of the infra-red data system just one month after first talking to Tryggvason about it and did a lab demonstration of how it could remotely record heart signals from students wearing special transmitters. "They could walk anywhere in the room and the signal would get across," Tryggvason said. "They could hide it under the table. It was quite amazing." Amazing enough, in fact, to shake loose some funding to carry on the project.

"They're all keen space nuts now, all four of them," Tryggvason added. And there was a spin-off effect on other students when they saw their classmates working on an exciting project and getting flights on the NASA plane. "That's how we have to get the really good ones. We've got to get the ones that are really bright, coming into university or coming out of high school — get them thinking about this so that when they decide what they're going to study at university, and especially in graduate school, they've already got this in their minds."

A bonus of these projects was that they also got the professors thinking microgravity, he said. "My goal is to get some of this stuff out to the universities . . . to ease people into gravity sciences. Part of the problem we'll have in attracting top scientists to work on space station is to get them into this mode of thinking what they can do using microgravity. If they're not thinking that way, they won't ever think of it, because they're busy doing their own thing, [and] they're not going to take the time to really educate themselves on microgravity. So we have to nudge them into that direction."

Thus, the KC-135 program, which was begun primarily to bridge the gap between shuttle flights, evolved into a useful program in its own right with long-term implications for Canada's scientific efforts aboard space station. It also helped the astronauts to feel genuinely productive and useful while they waited out the post-*Challenger* drought. "The KC-135 program has sort of saved us," Thirsk said in 1991, just before Bondar's mission got them back into flight. "It's about the only operational type of stuff we're doing." And with all the other scientific projects they undertook, "we've kept very busy. I think we can pat ourselves on the back that, although we haven't been flying, we have made a significant contribution to the manned space flight program in Canada."

Still, as 1991 drew to a close, the psychological wear and tear caused by the long delay seemed to be surfacing among the astronauts — possibly triggered by the fact that the two missions for which they'd originally been chosen those long years before were finally, tantalizingly, on the horizon and the fact that CSA was gearing up to bring new astronauts on board. After years of waiting, things seemed to be moving again — and, more importantly, changing — and this seemed to bring out a complex ambivalence and uncertainty in their reflections about what had gone before and what lay ahead for them. In November, a *Canadian Press* article appeared under the headline, "Astronauts disillusioned about calling," which, perhaps unfairly, captured only the downside of the sentiments expressed by the astronauts in the text of the story. While they openly acknowledged their frustration with the delays, MacLean, for example, also said, "I feel like I have the best job in the country."

There was also some ambivalence about the implications of bringing in a new group of astronauts. It was not that they were against the idea, particularly since they'd been hard pressed to meet all the demands placed on them. Nor did they believe this was an attempt to shunt them aside. "There's never been any indication that the space agency is looking to ease us out the door and replace us with new people," Tryggvason said. "It's never been hinted at." And Garneau viewed the hiring of a second generation as a positive sign about the future of the astronaut program. "It means there's continuity . . . and it gives you a feeling of perspective: you're no longer the only group, you're now the original group, and there's life after the first group."

But that didn't mean the first group was ready to be put on the shelf, and they felt it was essential to the morale of the entire team that the original astronauts still in the program, as well as the new ones, have reasonable expectations of getting a flight. "Our concern is [that] we have only three flights so far, and we do not have any firm flights in the future right now," Tryggvason said at the time. "You just take the total head count of astronauts, compare that to the total realistic flight opportunities, and get the numbers to match up, so that you're trying to fly the people you have in the program. We just felt it was premature to be hiring more people, We have to secure flight opportunities, [and] it has to be backed up with the money to fund the work for the flight. If they don't do that, then what's the point of hiring more people?" He compared the situation to running an airline without aircraft. "If I don't have any airplanes, why would I hire pilots?"

There was no denying that things had turned out very differently from what they'd expected when they first got into the game. While there'd been a lot of interesting scientific and engineering work to do, a lot of globe-trotting and interesting people to meet, and adventures of various sorts — much of which they could not have done anywhere else — there had also been a lot more desk-sitting, paper-shuffling, and bureaucratic hoop-jumping than they'd anticipated. And, of course, far less of the flying in space that had drawn them into the program in the first place.

"When I joined the program in early 1984, I expected that 90 percent of my job would be in astronaut training — the physical stuff as well as the academic type of training," said Thirsk at the time. "We were going to be flying soon, and there was a hint at the time that the astronaut program would become a long-term program, and I was naively thinking that we would be flying every year, that I would have flown twice by now. Of course, flight opportunities aren't what we dreamed they'd be. We still haven't flown the original two flights . . . and the nature of the job has changed. I'm doing work that an agency engineer or scientist would typically do — I'm getting involved in experimentation on my own, I'm monitoring experimentation that other people are doing for us under contract, I'm getting involved in committee work. All of which is very exciting, but it is certainly different than what I intended my job to be."

He acknowledged that the delays and the lack of flights had frustrated his career development. "You always want to have significant milestones that you can list on your CV . . . such as participation in space flight. The type of work we've been doing is important, but there's nothing that other people would interpret as meaty. I think I'm doing a decent job and making a contribution, but if I were to write up a CV and say that from 1983 to 1995, I was supporting the Canadian astronaut program, that doesn't say a whole lot." But in the next breath, he added, "I really enjoy the program even as it exists right now. We've been able to find fulfillment in the program even though the nature of it has changed because there are so many exciting things to be involved in. The manned space program in Canada is very young, and I sort of feel like I am a pioneer. I want to stay and shape it and help it grow. The level of current activity is nothing to brag about right now; it's something that has to be changed, and I'd like to be an agitator to get it changed."

Just after being chosen as an astronaut, Thirsk expressed a desire to stay in the program for "the rest of my life. If they want me for the next 20 years, I'll stay." However, by mid-1992, the delays and lost opportunities had taken their toll. After the *Challenger* accident, when Bondar and Money had been designated by the space agency for the life sciences flight, Thirsk said he was assured he would be good mission specialist material. "I got very excited about that, and I tried to guide my career development in that area." Then, in June 1992, Garneau and Hadfield were chosen for that assignment. Thirsk was philosophical about Hadfield's selection, acknowledging that his test-piloting qualifications were significant. "When it was decided to go with one new guy and one old guy, that hurt a bit in the sense that I have paid my dues. But on the other hand, if the reason for selecting Chris was that he was deemed to have more experience and to be a better candidate, I have no problem with that. I have always said we should not select people for flight for reasons other than technical merit."

He could not help but view this as a personal watershed, however. The eight years he'd spent in the program had been "an interesting and exciting time for me, because I was doing the ground-based work, but that's not enough. The reason I joined this agency was to fly in space. That carrot is being dangled in front of me, and I haven't been able to partake in it yet. If someone had told me back in 1983 that I would not fly for twelve years — the eight years I've already put in the program, another four to fly — I probably would have turned down the opportunity. The sacrifices my family has had to make for me have been more than I would ever have asked them to — the pressure on me, the salary hit I've taken, the time spent away from not only my children and my wife, but also my household responsibilities."

His career, too, has been affected. "My job description hasn't changed, whereas all my professional peers are moving up the ladder." In fact, Thirsk said he could empathize with Rob

Stewart's last-minute retreat from joining the astronaut program. "Everyone thinks it's all positive, but that's not the case."

Thirsk admits that friends and family have urged him to "get
out and cut [my] losses," but he feels that if he does so, "I would
be looked upon as a failure. I know the reason I haven't had a
flight to date is not because of inability, but because of circumstances. So I know I'm not a failure and my close family knows
that, but if I were to leave now, the media would perceive that
I was a failure, the public and my professional peers would
think that. If I could get a flight within three to four years, I will
stay in and tough it out, but if I wouldn't fly for at least eight
years. I would have to get out." That, for him, marks the
balance point of "ego versus practicality."

To improve his chances of getting a flight, he wants to spend
more of his time developing operational skills — for example,
"getting involved in human-machine and human-computer
interface design work on the mobile servicing system and
other pieces of hardware. I want to get more up-to-date on
some of the shuttle systems and start looking at some Canadian payloads that I might potentially fly on the shuttle. I'm
very frustrated right now because of the lack of career development, but I still have hope. I'm an optimist, [and] that
component of my makeup is keeping me in the program. I
think the chances are better than not that I'll have a flight
within four years from now."

For Bjarni Tryggvason and Ken Money, much of the post-
Challenger hiatus was devoted to activities associated with
being backup astronauts for MacLean's and Bondar's flights.
Like Thirsk, they knew they were spending a lot of time on
missions they probably would not fly. Nor did either of them
see any obvious prospects for future flights. In fact, the delays
had put Tryggvason in an usually awkward position — while
tied to an assignment that would in all likelihood leave him on
the ground, he missed the chance to take the mission specialist training course that was much more likely to culminate in
a flight. Unfortunately, the timing of the two events was bad
for him: the mission specialist program began in August 1992,
just as the training for MacLean's flight entered its final three
months. Although he much preferred the idea of being a mission

specialist, Tryggvason was nevertheless hoping to parlay the work he'd done on the SVS and his KC-135 projects into a payload specialist flight in a few years. "There is interest on NASA's part to fly the advanced space vision system," he said. He's also proposing a flight demonstration of the vibration isolation device he had developed for the KC-135. "Those two together could be ready for early '94. That's a reasonable flight for me to go for."

Whether CSA could find funding for the flight remained an open question. But Tryggvason said that even though he would be disappointed never to fly a mission, his future was nevertheless tied to the space program because of the research he'd done during the post-*Challenger* period. "I'm not going back to what I was doing before. I'm doing a lot of things that I'm enjoying, in terms of getting into microgravity research myself, [so] I'm going to work in the space area. All these little gadgets we're working on are getting to the point where I can do that."

Ken Money didn't regret the eight and a half years he spent in the astronaut program. He was, of course, disappointed at not getting a flight, which he described as "a major ambition . . . but not an obsession." Perhaps because he was older and had come into the program after having achieved considerable success in his own field, he was able to be philosophical about the experience. "It's like never having won a gold medal at the Olympics or never having won the Nobel Prize. There are all kinds of wonderful things you'd like to do and you're never going to do. This doesn't mean you haven't had a lot of wonderful opportunities anyway. I learned to speak French, I learned to skydive, I learned to fly a couple of new airplanes, I learned an awful lot of science from a broad area. It was a very enriching and exciting experience. I did not win the gold medal, but nevertheless feel that I have been at the Olympics."

Money, who'd been flying since he was 18, was already an accomplished pilot, qualified on jets and helicopters, and once flew search and rescue missions with the Canadian Forces Air Reserve. As an astronaut, he added two aerobatic planes, a Pitt Special biplane and a Super Decathalon, to his log book. He also took to jumping out of planes with great enthusiasm. "I really loved it," he said of his skydiving experiences. (Poking mild fun at the long delay in space missions, the ever-humorous Money scribbled on a photo of himself dropping through the sky, "A guy has to do his best in the absence of shuttles.") Since he'd been a long-time pilot, his wife, Shiela, wasn't too surprised to find him doing loop-the-loops in the aerobatic planes. "Anything with airplanes, that was just my usual," he said. "Skydiving was a little different; she found it a bit worrisome. She didn't complain about it, but she refused to watch. I said to her, 'You ought to come out and see me do it. It's really fun. You drop down out of the sky and it's fantastic.' She said, 'You're crazy. You screw up and splat! Do you think I want to be there to see it?'" She was no doubt relieved when Money gave up skydiving after becoming the backup payload specialist for Bondar's flight because of a NASA policy that forbids astronauts in training for a shuttle mission to engage in risky hobbies.

After the flight, Money resigned as an astronaut and returned to his lab at DCIEM, but he retained close ties to the program. He taught a space medicine course to the new recruits and remained available for occasional public appearances and speaking engagements. And he stayed on as CSA's representative at the International Academy of Astronautics, becoming chairman of the human factors committee for the Mars mission study. "I can't complain," he said when it was all over. "I've certainly had my share of goodies, and my experience in the Canadian astronaut program was enormously rewarding."

Medicine got Roberta Bondar through the waiting. "I love neurology, and I really enjoyed doing my stroke research. I've been able to compensate for any sense of frustration." Very much a scientist, she worried about keeping her hand in as a researcher. She knew that on her shuttle flight, she would be doing other scientists' experiments (when it comes right down to it, the job of a payload specialist can sometimes seem to be that of a highly overqualified lab technician). "You can't be too creative when you're dealing with black-and-white procedures," she said. "You can't really have an effect on the science being studied; that's everyone else's prerogative."

For Bondar, this was not enough — she needed to feel that she was continuing to make a contribution as a scientist in her own right. "I've enjoyed the work, but if I can't temper that with doing things in my own area, it's not as rewarding for me to be doing other people's science day in and day out. It's one thing to be known as the second Canadian in space or the first woman or the first neurologist or whatever they want to call me, but that will only have its fulfillment at the time of doing it and shortly thereafter. I have the rest of my life to live. I don't want to lose sight of the fact that this is not the be-all and end-all to my life — though I must say it is difficult."

She felt it was important for her professional future to continue doing research that would help preserve her skills in earth-based medicine while enhancing her space-related expertise. Not only did she believe that her job as an astronaut required her to stay on top of research in her field, she also had to consider "what would happen if there was another accident in the [shuttle] program" that might cause further flight delays. "You have to be a bit more firm about the kinds of things you do with your life. People keep asking me what I am going to do in the future and I can't answer that, but what I'm doing right now are little stepping stones that will leave doors open."

Steve MacLean was in a somewhat different position than the rest of the astronauts. He'd been assigned to his flight in 1985, before the *Challenger* disaster happened, and remained the only one with a sure flight opportunity for a long time afterward. (Although Bondar and Money knew they were in the running for the life sciences mission from around mid-1986, it was not until early 1991 that Bondar was selected as the prime crew member and Money as the backup.) MacLean found, however, that this did not make the delays easier to cope with. "I would be dishonest to say they're fine," he acknowledged. "Access to space has changed, and my career has changed because of the [*Challenger*] accident. The time frame for things happening is much different than I expected; I was to fly in '87 . . . and I'm flying in '92. Though we've done a lot in that time, in many ways it's not measurable."

When he was selected as an astronaut, MacLean was a

postdoctoral student at Stanford University in California, planning an academic research career in laser physics. It wasn't easy for him to give up that dream — "it took me a long time to make the decision," he said — but he saw the space program as an opportunity to combine his research ambitions with a childhood fantasy of becoming an astronaut. But the wide variety of duties expected of an astronaut meant that he couldn't spend as much time on his scientific research as he might have in an academic setting, and he admits he found the bureaucratic aspects of the work frustrating. And he worried about hanging on to his sense of curiosity and wonder — "the things that really turn you on in science, that's what got you into this in the first place. When you're a kid, you see something in nature and you try to figure out how it works and it gets you really excited. Sometimes we don't get enough of that. You have to sort of discipline yourself to keep your skills fine-tuned. But I don't have any regrets. I'm very excited to be here, and I feel so lucky that I'm going to fly. I think this is such an exciting program; it's worth it for me to hang in there."

MacLean was fortunate in one way — his work on the space vision system gave him an opportunity to maintain his interest in lasers and their application to space operations, something he hopes to continue after his flying days are over. "I think I will always be connected with space somehow. My expertise lies in laser physics — there's no question about that for me. If I thought the laser program would be a certain percentage of the Canadian space station program, I could see really getting involved in that area." His thoughts have even moved on to Mars and the possibility that lasers and robotic machine vision systems will be used in exploring the red planet. "We had all kinds of great ideas for Mars, and that's exciting — you'd need a land rover on Mars and you'd need vision. We can solve some of the problems there."

Marc Garneau's perspective on the delay was unique, since he was the only one of the six with a flight already under his belt. He stayed in the program partly because he was hoping for another flight, but also because, as he put it, "I owe the system something — that's why I've been around since 1984. If we all fly and then disappear, there's no continuity. I want to stay on, to pass on whatever I have to offer. If I had the opportunity to fly again, I would definitely not leave the day after." (Later, after being selected for mission specialist training, he reaffirmed this intention. "I think I've given back to the program in the last eight years what it gave me to fly as a payload specialist. If I left tomorrow, I wouldn't feel guilty. However, with the mission specialist training, I will feel a similar requirement to not go down for four years and then disappear, but to impart my experience when I return.")

The passage of the years had started to dim the memories of his shuttle flight. "It is beginning to fade. You can't help it. The essence will always be with me, but time is passing. . . ." And it was disconcerting to find he was sometimes viewed as the grand old man of the Canadian astronaut program, an "old and respectable type" who's had his day. Just into his 40s, he definitely had his eye on at least one more shuttle flight. But the delay had taken a toll, he admitted; when people asked if he was still an active astronaut, he replied: "I'm not sure sometimes."

In late 1991, however, he ventured the opinion that the morale of the astronauts had never dipped too low. "You have to be a dreamer and an optimist to be in this job. [And] the potential payoff is very great. Obviously there was big disappointment, and we were wondering whether things were going to pick up again after *Challenger*. Now everybody's working right up against the wall to get ready, and hopefully we'll have two successful missions [in 1992]. And we're wondering what's going to happen beyond that. We're very glad we're exercising the mission specialist option; we pushed perhaps harder than anybody else to get this. And we're glad that . . . we will hire new people and things will move on."

3

Flying Again

How many Canadians do you think I'm carrying on my shoul-ders? But they don't weigh anything in space . . .
— ROBERTA BONDAR

When Roberta Bondar emerged from the astronaut crew quarters at the Kennedy Space Center in the pre-dawn hours of January 22, 1992, she looked as though she'd already gone into orbit.

Never mind that her space shuttle was still sitting on the launch pad several kilometres away and that it would be hours before *Discovery* actually carried her into space — she was soaring. Dressed in a bulky orange launch suit festooned with colourful patches, glasses perched on her nose and a big grin on her face, she raised both hands in the air and called out, "Yes, yes, yes," to the clutch of reporters waiting near the van that would carry her and her crewmates to the launch site.

Later, in response to a comment about her ebullience that morning, she joked, "You can't peak too soon." It was a wry reference to the more than eight years of preparation and the 19 postponements that she'd endured waiting for the flight, at times wondering what it was all for. The grin and the high-fives, she said, were mainly for the benefit of the 65 family members and friends who'd come to Florida to send her off. "I really wanted some contact with my family and all my friends who'd gathered at the launch. My family was not there at the transport van, [but] I knew they'd be seeing it on TV. I wanted them to know that I was all right, to relieve their anxiety a bit — that this was going to be a lot of fun, and I was going to get on with it after all this training. It was a very tiring training program, and it was a great relief to feel we were getting it done."

"Getting it done" entailed eight days of intense work on what was probably the most demanding scientific research mission ever conducted on the space shuttle. Working in the European-built Spacelab located in the shuttle's cargo bay, Bondar and her six crewmates performed 42 experiments on behalf of more than 200 researchers from 13 countries. The first of several planned International Microgravity Lab (IML) missions, it focussed on how humans, animals, plants, and materials react to and behave in the space environment. The objective of these experiments was to study the effects of microgravity and cosmic radiation on biological samples, including fruit flies, frog's eggs and sperm, roundworms, mouse kidney cells, wheat seedlings, and slime mould.

The humans on board were also experimental subjects. A number of experiments, including several developed by Canadian researchers, studied the effects of microgravity on human physiology and performance. For example, the "Mental Workload and Performance Experiment" required astronauts to use an adjustable work surface and a portable computer to evaluate the design and efficiency of work stations in weightlessness.

Canada's contribution to the mission, in addition to Bondar's participation, was a group of six experiments, some of which continued research done on Marc Garneau's 1984 flight. Developed by 14 Canadian and 9 U.S. investigators, they included:

Space Adaptation Syndrome Experiments (SASE). When astronauts first get into space, they often experience a variety of symptoms, including nausea and vomiting, visual illusions, and a reduced ability to sense the position of their arms and legs. Researchers believe these phenomena occur because, as the human nervous system undergoes a complex process of adapting to weightlessness, the brain receives conflicting information about body position and movement from the eyes, the balancing system located in the inner ear (the gravity-sensing otolith and the rotation-sensing semicircular canals), and gravity-sensing systems in the muscles, tendons, and joints (the "proprioceptive" sense.) "Your body changes to become really a different being when you're in space," says Bondar.

This experiment involved several different tests designed to measure changes in the performance and behaviour of these physiological systems in weightlessness. It also included a test of the touch sensitivity of fingers and toes — a measure of whether tactile acuity has been affected by a nerve block caused by the stretching of the spine in microgravity. The head of the research team for the SASE experiment was space physiologist Douglas Watt, director of McGill University's Aerospace Medical Research Unit, who also designed a small space sled that was used for a number of the tests.

The goal of this research is not only to improve the understanding and treatment of motion sickness in space, but the diagnosis of inner-ear disorders and motion sickness on earth.

*N*ystagmus. The balancing system of the inner ear plays an important role in maintaining focussed vision by relaying information to the eyes about the movement of the head. If the inner ear is damaged, these signals are no longer sent, and the eyes may experience oscillations, known as nystagmus, that cause dizziness and blurred vision. There are two types of nystagmus, one in which the eye oscillates at about the same rate regardless of head movements and the other in which the oscillation rate is affected by the position of the head. Gravity is the determining factor in the latter case, so the experiments done in space will help researchers improve the use of nystagmus as a tool in diagnosing inner ear disorders on earth. The principal investigator on this experiment was Joseph McClure of the London Ear Clinic in London, Ontario.

*E*nergy Expenditure in Space. Physical fitness is as important for astronauts as it is for athletes. They need good fitness programs not only to keep up with the demands of training and their mission duties, but to counteract the negative effects of microgravity on muscles and bones. Designing proper dietary and fitness programs for space flight requires knowing how much energy astronauts expend while working in space, and this was the objective of an experiment designed by Howard Parsons and his colleagues at the University of Calgary's faculty of medicine. The group developed a simple technique for measuring energy expenditure that involves having the astronauts drink water enriched with variants of the oxygen and hydrogen atoms at the beginning and end of the mission and monitoring the concentrations of these atoms in urine samples collected daily during the flight. From these results, the scientists can calculate the amount of carbon dioxide produced, a measure of energy expenditure. This information will help scientists develop suitable diets to maintain body mass in space, which will become increasingly important as longer missions become more common.

The advantage of this technique is that it eliminates the need for the complex equipment used to measure energy expenditure on the ground. It is also being tested for use on earth to assist the recovery of trauma and surgery patients who cannot be monitored with the standard equipment.

*V*enous *Compliance and Experimental Anti-Gravity Suit.* Like the nervous system, the body's cardiovascular system also adapts to weightlessness. Without the pull of gravity, blood and other fluids shift higher in the body, triggering a response that ultimately results in a reduction in total blood volume. On returning to earth, the blood is pulled back into the legs, away from the upper body and the brain, and this could cause dizziness, faintness, and a loss of peripheral vision, particularly since blood volume has been reduced. To counteract these effects, returning astronauts wear inflatable anti-gravity suits that squeeze the lower body and force blood back up to the heart and brain. In this experiment, measurements were made before, during and after flight of the blood flow and volume in the veins of the astronauts' legs, to document any changes that occur in flight and gain a better understanding of how the veins adapt to weightlessness. Shortly after the shuttle returned to earth, one of the astronauts donned Bob

Thirsk's experimental anti-gravity suit, and the blood volume and flow tests were done again. The purpose was to compare the new suit with the conventional anti-gravity suit and with wearing no suit at all. Thirsk was the principal investigator on this experiment. After the flight, he said the preliminary results indicate the new design is "more effective than the conventional suit" but more research is needed.

*B*ack pain.   Astronauts often complain of back pain, particularly during their first few days in space. It was thought that this might be caused by the spreading apart of the spinal vertebrae that occurs in microgravity, which results in astronauts "growing" about five to seven centimetres. In the hope of finding ways of alleviating this problem, Peter Wing, a back specialist with the University Hospital of the University of British Columbia, developed assessment procedures that the astronauts followed in space. In addition to keeping a daily record of the location of any pain they experienced, they also took photographs of each other's backs to record physical changes. Similar pictures were taken after the mission, and the astronauts also received back examinations and x-rays.

*P*hase Partitioning Experiment.   This was not a physiological experiment in the same sense as the others. Instead, it was a "materials-processing" experiment that exploited the microgravity environment to separate out different components or "phases" of a complex mixture of substances. Phase partitioning can be used to separate biological substances used in medical research and treatments, such as bone marrow and pancreatic cells. However, on earth, it's difficult to achieve a high degree of purity in the separated substances because of gravitational effects on the fluids. This experiment, developed by Donald Brooks of the University of British Columbia's department of pathology, and his colleagues, was designed to study how the absence of gravity changes the partitioning process. A follow-up experiment was performed by Steve MacLean on his flight. Although the technique is mainly of interest to medical researchers, it also has potential as a method of producing better alloys.

A number of other materials-processing experiments developed by international research teams were also part of the scientific program on the IML mission. They focussed on the potential of using the microgravity environment to improve the processing of new pharmaceuticals and biological products and electronic materials for use in computers, lasers, and other high-tech equipment. One day, it may be economically feasible to produce some of these materials in space, but in the meantime, researchers have found they can often improve ground-based production methods after gaining a better understanding of the fundamental processes involved by eliminating the complex gravitational effects that occur on earth.

As a payload specialist, Bondar's primary responsibility was performing these experiments on behalf of scientists from Canada and the other participating countries. German astronaut Ulf Merbold shared this role, and the other crew members also pitched in, though they weren't required to. In fact, it's very unusual to have NASA pilots and mission specialists volunteering to participate in the scientific experiments done by the payload specialists, especially physiology tests. "Not everybody likes to be a guinea pig," said Alan Mortimer, chief of space life sciences for the Canadian Space Agency, who served on the ground support team at the Marshall Space Flight Center in Huntsville, Alabama. "But we had all seven crew members participating in the Canadian experiments. They were very helpful, and it allowed extra scientific data to be obtained."

Ken Money, whose job was to communicate directly with the crew on behalf of all the scientific investigators, said that the spirit of cooperation actually began long before the flight. "Months ahead of time, they were training with us to learn how to do various things so they could help." This was largely because of the attitude of commander Ron Grabe, who was "unlike almost all the other orbiter commanders" in encouraging such activities. "He's a really nice, sensible guy and thought it was a reasonable thing to do. I don't know how he had the nerve to do it. Most orbiter crews will not have anything to do with the working payloads." He said this is partly a reflection of a desire on the part of many astronauts with operational responsibilities to "come home and be able to say that not a single mistake was made. There is a tendency even to reduce the productivity of a mission in favour of having a mission in which no mistake would be made. I guess science tends not to work like that. You tend to try to maximize the return. If you make a mistake, that's a pity, but you nonetheless want to get as much done as you possibly can."

In the view of many of the scientists who worked on the mission, it more than fulfilled that goal. Shortly after the flight, a scientific colleague told Money that it was "the most productive scientific mission they've ever had. In some cases, they got a hundred and ten percent of what they were expecting."

The work went on 24 hours a day, with crew members and their ground support teams divided into "red" and "blue" teams on 12-hour shifts. Bondar was on the blue (day) shift. Prior to the mission, the crew spent much of their time in rooms specially equipped with bright lights to reset their biological clocks to the schedule they would be keeping during the flight. Sleep researchers have found that carefully timed exposure to bright lights can cause a fairly rapid adjustment to a new time schedule, and they are investigating the use of this technique to help people adjust to shift work and jet lag. The flip side of the coin is to avoid light exposure at the wrong time, so astronaut crews sometimes also wear dark goggles when they're outside in the bright sun. Along with her crewmates, Bondar spent most of her time in a desperate race to cram into seven days a series of experiments originally designed for a 10-day mission. (In the end, they got a much-needed extra day in orbit.)

The pace was made all the more gruelling by the fact that it often takes longer to do things in microgravity than it does on earth. "Work is inefficient because you're floating around and having to trap things," Bondar said. "You try working in an office with stuff floating around. It takes extra time to secure everything." This is one reason why duct tape and velcro are popular items on the shuttle. There were about 1000 scientific items aboard Spacelab that had to be kept track of, but they weren't the only problem. "You also have to deal with the stuff you're trying to throw away floating around," she said. Remarkably, Spacelab didn't have a garbage bin so the crew had to resort to using plastic bags stuck down with duct tape. "We need fly paper up there," Bondar said exasperatedly. "We need one of those creatures from Star Wars who gobbles the trash."

The lockers were another unending source of adventure. On earth, the doors to the overhead lockers drop down easily because of gravity, but in space, Bondar says, "they're the devil." So are the "exploding lockers." Packed tightly with clothing in earth's gravity, in space, "if you pull out one sock, everything comes out like a jack in the box — sproing!" Good luck trying to cram it all back in. At least on earth, Bondar says, "you can sit on the suitcase." In microgravity, forget it. (Of course, whatever disappears into nooks and crannies during the flight comes back with a vengeance during the return to earth, and the astronauts are bombarded with projectiles raining down from their microgravity hiding places. "There were bits and pieces of garbage and stuff falling off the ceiling," said Bondar. "You're so used to seeing it up there that it's kind of strange to see it on the floor." It's a good thing they wear crash helmets during re-entry, she added.)

A similar problem arose with putting out the garbage, which went into a hole in the floor of the orbiter. "You had to stick your arm in and poke the stuff down," she said. "It would forever have a life of its own and come up, and towards the end of the flight it was pretty gruesome." And it sounded like the astronauts should get hazardous duty pay at mealtimes. Bondar said that it was not unusual to suddenly have a "ball of someone else's food" hit her in the face. It sneaks up in microgravity, so she sometimes didn't have enough warning to blink and got "zapped" on the eyeballs. One day, a dollop of shrimp cocktail got her that way. "So there were always those nagging day-to-day logistics problems that you cannot simulate on earth. It's really tough to prepare yourself; you try to think about these problems, but you never really know what they are till you get there."

In fact, her schedule was so jam-packed that Bondar could barely find time to eat. "She worked through her lunch hours routinely," Money said. "The medical guys in Houston were saying, 'You know, she really should take a break. She shouldn't be working like this.' They tried to say, 'It's your break now. Lunchtime. Would you go to lunch now?' That happened twice, where I strongly suggested that she take a lunch break."

It wasn't that Bondar wasn't interested in food. After the mission, she commented that she felt very hungry — the astronauts generally feel the "food containers aren't big enough and the servings are small" — and she took to carrying snacks around with her because the food was so hard to get at. But with her heavy workload, she sometimes felt she had to choose between food and sleep. "I knew if I took my lunch break, it meant I was going to be later getting to bed and putting the timeline further behind."

It was like that for the others, too, said Alan Mortimer. One of the U.S. astronauts, Norm Thagard, also regularly worked through his lunch break and exercise period. Sometimes astronauts would go off duty without having finished some of the experiments, and the ground support team wouldn't find out until the next day that they'd either taken work away with them or quietly finished it in the lab while officially off duty.

It didn't help that it was hard to get to sleep because the shuttle was so cold at night, said Bondar. The sleeping area was located against a wall of the spacecraft that faced deep space in order to dissipate heat that built up on the vehicle from the sun. Since the temperature outside that wall was near -273°C, the cabin got "quite chilly," said Bondar, who found herself bundling up in double sets of clothing and winter-weight long johns. "I didn't have any gloves, but if I had, I would have worn them." She even stuffed clothing in a cold air jet that was intended to blow carbon dioxide away from the sleeper's face, joking that she reached the point of preferring suffocation over freezing to death.

The intense pace of the mission was shared by the scientific ground support team, which was camped out in what Alan Mortimer dubbed "the dungeon," a floor below the scientific mission control room at the Marshall Space Flight Center in Huntsville, Alabama. (The primary mission controllers were, of course, at the Johnson Space Center in Houston.)

Each of the experiment teams had a table with television sets and communications consoles that allowed them to observe what was going on in Spacelab and communicate with mission controllers. Most of the Canadian scientific investigators were there. "I had a team of 15 on the ground," said Mortimer. "We had everybody there, and they were all working when their experiments were being performed."

As one of the leaders of the Canadian ground support team, along with Bruce Aikenhead, Mortimer's job was "essentially to run interference for the scientists and make sure that things worked out so their experiments got done. We had stretches where there were a great many activities going on, one right after the other. The mission was very crowded and running late at times, and a lot of tension would build up. Everybody is under stress and everybody is tired — the crew was sleep-deprived the whole time — and the folks on the ground are working really weird hours, so by the end they get tired as well. So you're in a position of having to try to mediate."

At times, a bit of amateur psychology was needed. "You'd be sitting there looking at your watch and trying to guess whether you needed to intervene so that they got the opportunity to do all the experiments when they were supposed to, or whether the crew was in a good mood that day and would be willing to work overtime, so you could leave everything alone. There was a case late in the mission where we were sure that the crew must have got up on the wrong side of bed. A communication problem arose; the fax machine went down and we couldn't send a new schedule, and they misinterpreted what they got. The shift going off was tired and didn't tell them everything, and the shift coming on was tired and looked at what they were supposed to be doing and got very grumpy until we were able to get the communication problem straightened out." (At one point in the mission, Norm Thagard complained about the work load, reminding the mission controllers that the crew members were human beings, not machines.)

The scientific investigators were also required to respond almost instantaneously to requests for information or instruction from the crew. Even seemingly trivial matters could make a serious difference to an experiment. At one point, for example, a crew member called down to say there was only one bag left for collecting urine samples and asked whether he was the one they wanted to provide the sample. "He sort of needed to know right about then or he would lose the target of opportunity," Mortimer said.

These messages were being relayed through Ken Money and Roger Crouch, who, like Money, had also trained as a backup payload specialist for the flight. This proved to be an advantage, Money said, because they understood the experiments and procedures being done in space intimately, having practised them themselves countless times over the past year. "My job was to see that the work got done and that time wasn't wasted. If they needed information, I would get it for them. I had push-button access to all kinds of people." On a mission that was costing around $17,000 a minute, every bit of time saved made a difference.

Mortimer said that the scientists found it extremely helpful to have such a direct line of communication with the crew, something that doesn't usually happen. "On Marc's mission, you would sit there for days and wonder what was happening, whereas you usually knew what was going on in Spacelab. If you saw something you didn't think was right, it was quite easy for you to react."

Viewing himself a kind of science officer for the flight, Money dubbed himself SPOC, a whimsical acronym standing for "Spacelab Payload Operations Control" that he made up himself because he liked the sound of it. He said that Roger Crouch thought "it was a great thing to call us."

Mortimer said that the system worked so well that it will undoubtedly become the benchmark for coordinating complex international space flights. "By the end of the mission, you had an international group that was actually quite cohesive — a hundred and twenty-odd people actually able to make decisions that affected everybody in a fairly reasonable manner." He said the working scientists were able to "understand and sympathize with each other's problems. [Some] actually gave things up so that somebody else could accomplish more of their goals. In a number of cases, people said, 'We don't need that as badly as you need this.' It was one of the really pleasant things, and a bit of a surprise, how easily a large international group was able to work together."

This, no doubt, will also provide a model for space station missions, he added. "I would expect that space station is not going to be much different in the early stages . . . and I think IML-1 demonstrated that it does work." With this in mind, one of his major objectives is to find a way to maintain the expertise Canadians gained while working on "the other side of the microphone." Canada will be providing ground support teams as well as astronauts for the space station. "A great deal of the support we're providing, in manpower, is going to be in the ground and operational support for the MSS and operational support for science. You have to get that right, too. There are specialized skills required to do that, and we're trying to identify and maintain those skills."

Bondar's mission was not all work, however. There were lighter moments, such as the Great Coin Toss, in which Bondar played a starring role. The Superbowl occurred while the flight was in progress, and during a live television broadcast, Bondar, clutching a coin in her hand, was flipped by two of her crewmates. "I really wanted to flip the big guy, but there wasn't enough room," she said. She regarded the incident as nothing more than an effort to "lighten up" a mission that was otherwise very intense. "We paid for doing that. We ended up having to spend extra time to do the experiments we should have been doing at that time."

Like virtually all other astronauts, Bondar was fascinated with watching the earth go by below and stole as much time at the windows as the busy schedule allowed. *Discovery* was flying at a high enough angle to the earth's equator to pass over parts of Canada, and Ken Money recounted that "at one point, Roberta suddenly took off and flew at high speed a long distance to the back [of Spacelab]." Everyone on the ground was startled until Bondar explained — the shuttle was passing over her home town, Sault Ste. Marie, and she wanted to get to the window. Because the shuttle zips along at about 25,000 kilometres per hour, "if you don't get there smartly, you're going to pass what you want to see," Money said. "She said, 'I can see right across Lake Superior. You can see for a thousand miles. It's wonderful.'"

For Bondar, viewing the earth from space had the emotional impact of meeting someone "in real life rather than in a photograph." The experience obviously made a deep impression, because it was one of the things she talked about most ethusiastically when she got back. In a CBC radio interview, she commented that it gave her mixed feelings to see such dramatic evidence of both the earth's beauty and its isolation. "When you look out beyond, there's nothing else there that we can see. Nothing. There's no *Star Trek* vehicle that's going to bring a whale back to our oceans." Consequently, she came back with a greater appreciation for the uniqueness of everything on earth — including things she'd found to be "pests" before her flight.

On January 30, 1992, eight days after the flight began, the shuttle returned to earth. Re-entry was also a unique experience, Bondar said. "When we got down through about two-tenths of a g, I could really feel my arms become very heavy. Things just get progressively heavier and heavier." Being back on the ground feels "very, very strange," she said. "Your whole body feels different; you have the feeling that you're going to fall over very easily if someone pushed you. You can't walk a straight line for three or four days; you can't really maintain your balance. I found it most unpleasant. No one finds coming back very pleasant. You think about what would happen in an emergency if you had to get out, so people try to get their vestibular systems back as fast as possible."

Reflecting afterward on the intensity of the flight, Bondar commented, "I'm not sure NASA will try to put a flight together that's that complex again." And she concluded that space engineers still have a long way to go to create a truly efficient microgravity workplace. "The shuttle never was meant for long-term space flight. We're making do; I don't feel that we're as efficient as we could be. This raises the fundamental issue of the appropriateness of having a space vehicle and a training program designed around the one-g mindset."

The placement of equipment racks in Spacelab was just one example: "We train in laboratories that have up-and-down racks, the way we work in an office, and that becomes totally disorienting up in space." For example, during the first few hours in orbit, astronauts have difficulty recognizing which surface is which, and this can create problems when they try to do experiments rehearsed in one-g. Bondar said one of her crewmates preferred to use the computer workstation associated with the mental work load and performance experiment while "hanging" by his feet from a rail near the "ceiling" of Spacelab, rather than standing on the "floor" as he'd done in training on earth. Of course, in zero gravity, he didn't feel as though he was hanging upside down; in fact, he found the position "much more comfortable," Bondar said.

Bondar believes that NASA should be sending up experts in human factors analysis to experience these problems first-hand, so they can improve the design of future spacecraft. Each returning crew is questioned about these problems, and there's usually a thick file of suggestions, but few changes are made, she said. "We're really not getting people up there, I feel, who have the professional background to be able to fix some of this stuff."

As a medical doctor and neurologist, she's also interested in gaining a better understanding of the physiological and psychological factors that affect human performance in space, particularly during the early part of the flight when the human body is still adapting to the strange new environment of microgravity. On her flight, Bondar did an experiment of her own devising, which involved the use of a technique called transcranial doppler to measure blood flow in the brains of the astronauts. She is particularly intrigued with the question of the early effects of microgravity on orientation, memory, and thought processes. "You really don't think as clearly when you first get up in space," she says.

B ondar's pace did not slow down much in the months after her flight. After a period of debriefing with scientists and space agency officials at the Johnson Space Center, and a series of medical examinations, she returned to Canada in mid-February to begin a hectic schedule of public appearances, which included tossing out the first pitch at a Blue Jays baseball game, appearances at the United Nations and the summer Olympics in Barcelona, making speeches, and participating in press conferences. She was named Woman of the Year by *Chatelaine* magazine and one of 12 Canadians "who make a difference" by *Maclean's* magazine. One of her activities generated some controversy — she was one of a number of prominent Canadians featured in a series of TV ads run by the government extolling the virtues of Canada in honour of the country's 125th anniversary. There was no overt political content in what she said — she merely expressed the personal emotional impact of seeing Canada from space — but the ad ran just as the Constitutional debate was overheating again, and the country was gearing up for a pitched battle over the referendum to decide whether to accept the Charlottetown accord. It was almost inevitable, therefore, that the ad would elicit comment. For example, Maude Barlow, a women's rights advocate and political activist, said, "What we need is not a bunch of astronauts and hockey stars telling us to go out and love our country. I want the text [of the accord.]" And a newspaper article about the millions of dollars the federal government had spent on the advertising quoted political critics disparaging these "feel-good campaigns" as an effort by the government to massage the voting public in anticipation of an election.

But Bondar denied that she had been cynically used for political purposes; she said no one forced her to make the ad, she wasn't paid for it, and she wasn't given a script. "It was all my own words, my own thoughts. It was totally spontaneous. I was asked if I'd like to say something about my pride in Canada. What am I supposed to say — that I don't like Canada? I like Canada. Is there something wrong with saying I like Canada? Very few people can get up and say, 'I saw my country from space.' Unfortunately, some people are going to confuse that with an endorsement of the government. It would have been different if I'd been asked to comment on the referendum."

Another controversial incident occurred before the flight, with the publication of a photograph showing Bondar, dressed as a Mountie, and Norm Thagard, dressed as a Seminole

The first group of Canadian astronauts. Back: Ken Money (left), Roberta Bondar, and Bjarni Tryggvason. Front: Bob Thirsk (left), Steve MacLean, and Marc Garneau. (CSA)

Dave Williams got the call while teaching at the Sunnybrook Health Science Centre in Toronto. He was jumping up and down — mentally — but didn't miss a beat as he resumed his lecture. (CSA)

Julie Payette, a Bell-Northern computer engineer, couldn't believe she kept making the cuts. (CSA)

Chris Hadfield, a test pilot, was stationed at the U.S. Naval Air Test Center in Maryland when he was called with the good news he had been chosen. (CSA)

Rob Stewart celebrated his selection with a night out in Paris, where he was attending a science conference. His sense of commitment to the University of Calgary and his students caused him to reconsider the chance of a lifetime. (CSA)

Mike MacKay was at the top of the waiting list — next in line should any of the chosen few drop out for any reason. (CSA)

NASA's KC-135, used for weightlessness training, is less than affectionately called "The Vomit Comet." (NASA)

Bob Thirsk enjoys a moment of fun on the KC-135. (NASA)

Ken Money practising free fall. (Staff Photographer, Skydive Spaceland)

U.S. astronauts working in the Weightless Environment Training Facility (WETF) at the Johnson Space Center. Canadian astronauts will now be participating in such exercises as part of their mission specialist training. (NASA)

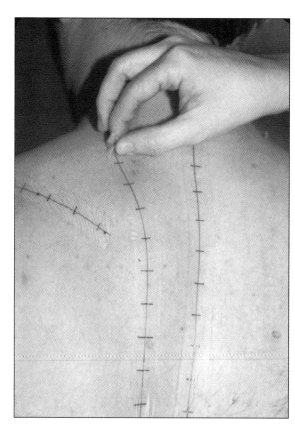

Back pain experiment on the shuttle. The markings pasted to the back are used to measure stretching of the spine in microgravity. (Dr. Peter Wing, UBC)

Roberta Bondar practises a space motion sickness experiment on the ground. (CSA)

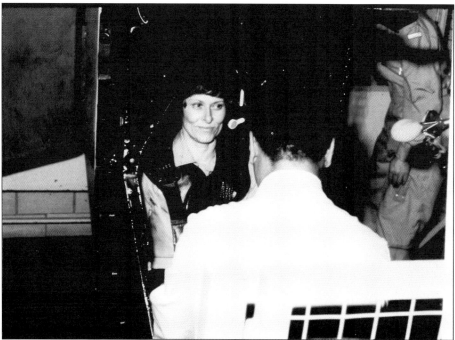

Marc Garneau at work in space. (NASA)

Roberta Bondar doing experiments in the International Microgravity Lab (IML) mission. (NASA)

Steve MacLean all suited up for his mission. (NASA)

Artist Paul Fjeld's view of Mobile Servicing System. (CSA)

Indian, standing back to back. A letter to the *Globe and Mail* suggested it looked more like a promotion for a wrestling match than for a scientific expedition. But Bondar said the picture wasn't meant to promote anything — it wasn't even intended to be publicly released. NASA tradition calls for shuttle crews to pose for informal, lighthearted pictures to be circulated among the trainers and other ground support people who work on a mission. Bondar said the mission commander suggested that since it was an international flight, she and Money should dress in costumes that were identifiably Canadian. Money chose the accoutrements of a hockey player, and Bondar became a Mountie. "I always wanted to be a Mountie, besides being an astronaut and a doctor, so I had my dream come true for a day," she said. Thagard's costume was described as representing the "mascot" of his alma mater, Florida State University. (A copy of the picture was sent to the university, and it was believed to have been leaked to the media from there.) Bondar dismissed the reaction to the picture. "Some people must have a lot of time on their hands to write things like this."

Over the years, she has become used to being the focal point of controversy and to having her every move scrutinized and commented upon. Inevitably, much of this comment centred on whether she was being favoured or discriminated against because she's a woman. CSA officials have always resolutely denied that this factor entered into their decision-making. But they've also made it clear that part of an astronaut's professional duty is to give the Canadian space program a high profile among the media and the public — indeed, the astronauts' qualifications for doing so were among the major criteria in their selection — and it's disingenuous to pretend that Bondar's status as the only female Canadian astronaut was irrelevant in this regard. The demand on her time for public appearances stems partly from the fact that she is widely perceived to be a role model for young women. Science minister William Winegard has said numerous times that more high-profile role models like Bondar are needed to encourage young Canadian women to take up scientific and engineering careers. "I can't believe this country doesn't have them, and I blatantly want to use them. I make no apology for that. We have to double the number of students in science and technology by the year 2000, and the only way you're going to accomplish that is if you can bring more women into these areas. There's no way in the world that you can get the participation rate of men significantly higher than it is; you've got to turn young women into these areas."

For her part, Bondar found it exasperating that the issue just would not go away, and she was frankly annoyed whenever it carried the taint of tokenism. (She was described in one Canadian Press story as making a retching sound to indicate how she'd react to having to answer one more question about what it was like to be a woman astronaut — but there were a lot more of those questions still to come.) No fan of quotas, she bristled at any suggestion that she was not getting ahead on merit when she'd virtually abandoned her scientific career and the rest of her life to a punishing schedule of travel, training, and public appearances, particularly in the three years leading up to her flight and for nearly a year afterwards.

One notable incident occurred in October 1990 during a visit to Ottawa by Valery Polyakov, then deputy director of the Institute of Biomedical Problems in Moscow, during which CSA officials announced the establishment of a joint Canadian-Soviet research program on the medical problems associated with long-duration space flight. Polyakov, a physician and a former cosmonaut who'd spent eight months aboard the *Mir* space station, commented at a press conference that Soviet space scientists were interested in studying the long-term physiological effects of weightlessness on women. They were lacking this information because they'd flown only two women cosmonauts, and neither had been in space longer than 12 days. He suggested that Bondar could give them advice "on how the conditions in space affect women" and said the possibility of having her fly on *Mir* was tantalizing because no women cosmonauts were scheduled to fly for several years (although consideration was being given to the possibility of flying an all-female *Mir* crew to collect such data).

A clue as to why long-term data on women cosmonauts was so lacking was perhaps contained in other remarks made by

Polyakov, who was then involved in selecting and training space crews. In news reports, he was quoted as saying through an interpreter that "if there is a woman on board, it creates an uneasy situation for the men. Long flights are hard work, both physically and psychologically — hard men's work." (No mention was made of the fact that male cosmonauts have sometimes experienced considerable emotional and psychological stress after spending long periods of time in space.)

An article in the Ottawa *Citizen*, headlined "Mir Machismo," highlighted Polyakov's disparagement of women. It described Polyakov as a "frontline fly-boy" who "says women are nothing but trouble." The article also quoted Polyakov referring to the "hysterical aspects" of women flying in space and said he had suggested that conditions in space were so harsh that it's impossible to predict how women would react.

In contrast, the *Globe and Mail* carried a more sedate headline — "Space Advice Sought: Woman astronaut could help Soviets" — reflecting the story's emphasis on Polyakov's ruminations about the need for more research on microgravity effects on women. The comment about women making men uneasy did not appear until halfway through the article, and nothing was said about women being hysterical or unpredictable. In keeping with the *Globe*'s editorial policy, both Bondar and Polyakov were given titles. However, while Polyakov was called "Dr.," Bondar, despite having a medical degree and a doctorate, was inaccurately called "Mrs."

As for Bondar, she was having none of it. She was quoted in a later *Globe* story as saying that she was offended by the idea of being invited to fly aboard *Mir* just to provide "the body of a woman" for physiological research. She had no quarrel with the objective of learning more about the response of women to microgravity and said she'd be willing to participate in physiological studies as part of a mission in which she was a legitimate scientific researcher and crew member. But she drew the line at being nothing more than a female guinea pig; others without her level of education and training could be found for that purpose. "I would tell them to get someone else."

In perhaps its most emotionally trying incarnation, the issue popped up again during the protracted competition between Bondar and Money to be chosen as the prime pay-load specialist for the International Microgravity Lab (IML) flight. Along with Bob Thirsk, they had come into the astronaut program knowing they'd be squaring off against one another for the Canadian life sciences mission. However, during the delay caused by the *Challenger* accident, the game plan changed.

Immediately after the accident, Thirsk, Money, and Bondar asked CSA officials to designate which two of the three would be assigned to the life sciences flight, so they could decide what to do about their careers. The assignment went to Bondar and Money. Although no official announcement was made at the time, it was later revealed in media reports, and subsequently confirmed by a CSA official, that the agency had selected Bondar to be the prime payload specialist and Money to be the backup.

It was not long afterwards, however, that CSA was invited by NASA to consider putting a payload specialist on the IML mission. Aikenhead said the opportunity offered some advantages over the original plan to fly the mid-deck experiments. "It would give us an opportunity for somewhat greater involvement in the mission activities, and it would also enable us to satisfy the desires of a fairly large group of Canadian investigators." On Garneau's flight, they had discovered the advantages of involving numerous researchers. "It was only as a result of the unexpected invitation for Marc to fly that we scrambled to find some other experiments and hit on what I think is a very successful scheme — to have a variety of experiments that . . . can cover a broad range of fields so we get a whole lot more out of our eight or nine days of flight. We had that in mind for both the life sciences mission and the SVS mission when they came along." Since Spacelab had more room and resources than the mid-deck, the IML mission would provide a much better opportunity to pursue this strategy. It would also give Canadian astronauts a valuable opportunity to work with researchers from all over the world — almost a practice run for learning how to function as part of an international space station mission.

So in early 1989, Canada threw in its lot with the IML

mission. It became a member of the Investigators Working Group (IWG) — the group of scientists developing the Spacelab experiments — and was given the right to have a payload specialist on the flight.

This changed things for Bondar and Money. No longer was the choice of the payload specialist up to CSA alone; NASA wanted two candidates put forth to participate in a year-long head-to-head competition that involved travelling all over Canada, the United States, and Europe to the labs of the IML scientific investigators, where they were briefed on the various experiments and were expected to demonstrate their competence to carry them out in space. In Money's words, they became full-time, travelling students for a year. U.S. payload specialist Roger Crouch and German Ulf Merbold were similarly pitted against each other. The idea was that the IML investigators, through the IWG, would ultimately decide on the two candidates they wanted as their proxies in space.

Early on, Money became concerned that "secondary" factors unrelated to scientific competence — primarily public relations factors — might play a significant role in the selection of the Canadian payload specialist. He felt, for example, that CSA officials would consider it a PR bonus that Bondar would be the first Canadian woman to fly in space. In March 1989, he expressed his concerns publicly on the television news show *W5*, which dubbed the competition between the two astronauts Canada's version of the space race. Money stated his belief that he had the scientific edge in the competition because two-thirds of the work to be done by the Canadian astronaut would consist of physiology experiments in which he'd been an acknowledged world expert for 30 years; during that time, he said, he'd worked with NASA scientists and the U.S. National Academy of Sciences on space physiology problems and had been a ground-based scientific investigator for experiments flown on five previous shuttle flights. He added that he'd published 80 papers in this field, while Bondar had published only three, and his papers were cited by other scientists 30 times a year compared to once for Bondar's papers. And although Bondar was also an established scientist, it was in a different field, he said.

Money was concerned that if public relations factors were given any great weight in the selection, the advantage associated with his scientific credentials would disappear and Bondar would win the competition.

A story published in the *Globe and Mail* around the same time had a similar theme. Money said he had not been able to "squeeze out" of CSA officials a list of the secondary criteria that would be applied to the selection process, but noted that he'd been told these criteria would rule him out and that Bondar would fly the mission. CSA vice-president Garry Lindberg was quoted as acknowledging that there were secondary factors — which he defined as "who can best represent the program" — but he would not elaborate further, except to state that he did not include sex as part of the criteria. The story described Money as sounding irritated when he compared the situation to a hockey match in which the winner is decided by a referee after the game, not by the number of goals scored.

Predictably, Bondar was far from pleased by any hint that she was being favoured because she's a woman. In fact, she vowed she'd quit if she believed that to be the case. And she emphatically rejected Money's contention that he was better qualified scientifically; as far as she was concerned, she was the better candidate because she had a diverse background that embraced more of the entire range of scientific fields covered by the IML experiments. She pointed out, for example, that in addition to her medical degree and her speciality in neurology, she also had a degree in agricultural science, which would be useful for the large number of the IML experiments involving plants and animals. (It is worth noting that neither Bondar nor Money was an expert in materials science, so both found it a challenge to get up to speed on the materials-processing experiments for which they were also responsible.)

"It was sheer nonsense that vestibular experiments were going to be the only thing being done on this flight," Bondar said in an interview about a year later, after she'd been selected for the flight. "There are 40-odd experiments in many fields, so I bristle when people say that because you haven't spent x

number of years in a certain science that you're not qualified to fly. We're looking at a broad base here. We're also looking at other skills." For example, she said the scientific investigators were looking for qualities in the payload specialists "outside the qualifications written on a piece of paper. This person has to be an extension of you, the scientist, in space, and their judgement has to be your judgement."

Asked what role being a woman had played, she said that she sometimes feels "I have to be overqualified to have my presence even noticed." But she was adamant that "affirmative action" had nothing to do with her selection. "I cannot, in my own heart, be involved in something like that. I believe affirmative action is fine at lower levels, to give people a foot in the door. Then what they do with their opportunity is their look-out. [But] it would be the worst thing to have a woman selected for this flight just because she's a woman, or because they thought it was good PR. That is sheer nonsense. The best person must be selected because otherwise you end up with egg on your face."

Bondar was reluctant at that point to rehash the emotional turmoil caused by the competition — "I have feelings about it that I don't want to discuss." In fact, she was reluctant even to characterize it as a competition. But she did concede that it had put the four payload specialists in "a very unusual situation. I think it was disadvantageous to have put us in such a position. I can understand why it was done, [but] no one liked it."

"It didn't have to go on that long," Money commented in a later interview. "They intended only to have a six-month period till they selected the finalists, but it went to 12 months."

The *W5* show did close on a conciliatory note. Money said whoever won would feel badly about denying the opportunity to the other. If he was chosen backup, he would do his best to provide Bondar with the support she needed, and he was sure she would do the same if their positions were reversed. In closing, the *W5* commentator observed that the one chosen as the backup might have to demonstrate "even more of the right stuff" than the one chosen to fly the mission.

Predictably, the TV show and the *Globe* article, both surprisingly gloves-off affairs, did not please the Canadian Space

Agency, which, like NASA, prefers its astronauts to present a happy-campers image to the public. But this kind of rivalry is both understandable and to some extent unavoidable, given the nature of astronauts, who are, after all, highly competitive achievers who've been used to success most of their lives. And, for the Canadian astronauts, the competitive situation was greatly exacerbated by the severe shortage of flight opportunities. This was only partly a consequence of the *Challenger* accident. Another factor was the lack of funds in Canada for bankrolling numerous shuttle missions and the ground-based research and development needed to support them. Also NASA was making it tougher for payload specialists to fly; the agency had to be convinced that a payload specialist had something worthwhile to do that their own astronauts couldn't do just as well.

NASA astronauts didn't have quite as many constraints, especially after the shuttle program got back on track. Not only were they eligible for far more crew positions — as many as half a dozen per flight — the increasing frequency of launches meant that one missed flight opportunity was not such a big deal; another would be along in a matter of months. Such was not the case for Canadian astronauts, to whom a missed flight opportunity could mean many more years of waiting and uncertainty, or, as in Ken Money's case, the end of a dream.

For Money, the irony of the situation was that, in the early 1980s when NASA first invited Canada to fly someone on the shuttle, he and Doug Watt proposed the physiological experiments with the idea that one of them would fly the mission. Both had had a long-standing involvement with the shuttle program as ground-based researchers, and they naturally assumed they would be the logical candidates for the job. But the Canadian government decided instead to parlay NASA's invitation into an opportunity to establish its own astronaut program, which ultimately led to the conundrum in which Money found himself a decade later.

The saga ended in almost as much controversy as it began, although the denouement was less public. For several months, there was much discussion and considerable confusion about the role of the Investigators Working Group (IWG).

Robert Snyder, NASA's mission scientist for the project, said one of the IWG's most important functions was to determine if anyone was "just not cutting the mustard." But that was not a problem. "They're all adequately qualified. Both Roberta and Ken are very satisfactory to the principal investigators I've talked to. They're both doing an outstanding job."

It was finally decided the IWG would hold a secret ballot to indicate which two of the four astronauts they wanted as prime payload specialists and which two would be backups. Snyder said at the time that an open vote was not a good idea because it would be "terribly disruptive. I really don't want anybody hurt by this." He also wanted to avoid the possibility that a scientific investigator might be later disadvantaged by having to work with an astronaut for whom he or she had not voted. "I certainly want to make sure there isn't going to be anything like 'He didn't vote for me so I'm not going to spend much time with his experiment.'"

The results of the vote were not made public at the IWG meeting. The reason given was that the choices had to be ratified by senior officials of NASA and the international partner agencies and publicly announced through NASA headquarters. "I'm afraid a lot of people are going to get their oar in this," said Snyder. "The IWG will definitely make a recommendation in some sort of priority order, but NASA headquarters or the Canadian government, anybody above the IWG, can decide to change that. That's always been the case." However, to forestall any uneasiness among the IWG that their selection might be overturned without their knowing it, a scrutineer from their group was present for the counting of the secret ballot, and he was free to inform the group if their choice had been overturned.

Finally, in early 1990, the IWG did vote, the scrutineer did not raise a fuss about it, and Bondar and Merbold emerged victorious. If politics played any role in the outcome, no one was admitting it — but the convoluted and secretive manner in which the exercise was conducted had done little to help dispel suspicions.

So far as Bondar was concerned, the issue was settled. "I had every confidence that the Investigators Working Group would see the strengths that each of us had and would make an appropriate decision without any political interference." She was informed of that decision during a phone call to her home by then CSA president Larkin Kerwin. "I was really happy that it was finally over," she said. "It was a relief. There had been a lot of strain — it was building right up to the last night. Things were happening that I really did not approve of and I was happy that the decision was made, and now let's move on and forget about it. Let's not rehash all this nonsense." But it never went away. The question came up again and again throughout her year of training and even up to the time of her flight. In a January 1992 article in *Homemaker's* magazine, she was still feeling compelled to say, "I was not selected for this job because I was a woman, no matter what the innuendos."

One thing did not change after the choice had been made: Bondar and Money were still students, still on the road more often than not, living out of suitcases and camping out in airports, waiting for flight connections to here, there, and everywhere. Not only did their training program demand regular jaunts between the NASA centres in Houston, Florida, and Alabama, they also continued visiting the labs of the principal investigators, learning more about the research and honing their ability to operate the equipment associated with the experiments. As the flight date approached, Bondar and Money started participating in simulations of the mission, some of which lasted for 36 to 48 hours at a time.

In addition, there were the physical demands of the safety training — jumping into the water encumbered with 40 kilograms of launch escape suit, enduring days of water survival training, tumbling out of the orbiter's side hatch windows, driving the armoured tank that would be used for an emergency escape from the vicinity of the launch pad. Bondar found this sort of thing fun, saying that she wished she could flash back to her past so her childhood self would know that her dream of going into space was coming true.

But there was a downside to all this furious activity, of course: Bondar had to put the rest of her life on hold. Starting about six months before flight, "we had to give them our total time — just go when they say go and travel when they say travel." The problem was that the launch date kept slipping, so this period of total commitment kept stretching, and her schedule was full of changes and uncertainty. The worst part, for her, was being unable schedule holidays and time off to spend with her family and friends. "Even going up to visit my Mom . . . you'd ask ahead of time, like a month or two months ahead, and they'd say, 'Oh yes, definitely we're going to have this week off.' So you tell all your family and friends we're going to have this big barbecue at the cottage, only to find out that that's been changed and you're not going to be up that week."

She added that there was little recognition of the needs of single people for contact with family and friends, which, in NASA argot, are known as "support systems."

"I don't have the excuse of having a husband or children," Bondar said, in reference to the comment made by one person with whom she was working who said that "after a spouse and children, the next level doesn't really count for much if you're trying to make an argument that you need to have some time with support systems. I think there should be time for support systems for single people, whether they are single men or single women." After her flight, she was upset that her mother was not accorded the same right of access as the spouses and children of the other crew members. Money noted that NASA has a rule that "only spouses can see their fliers at certain times and places, and [it] didn't make provision for people who didn't have spouses."

To keep up with the physical demands of the training and travel, both Bondar and Money managed to squeeze a regular exercise regimen into their tight schedules. But all the running around took its inevitable toll. At their final pre-mission press conference in November 1991, they both were described as looking very tired.

And during that press conference, Bondar was still trying to lay the "woman thing" to rest. When she was asked about the significance of being the first Canadian woman in space, she said: "I think that it is a message to Canadians that we have Canadians who are able to do a job regardless of what their sex is . . . I would like to think that one of the most important things I'll be doing . . . is the fact for all people to see that a woman is as capable of doing this job as a male and that there is not a difference. It is a personal attribute rather than one of gender. I was brought up as a person; I wasn't brought up [with] a male or female attitude towards life. I just feel that you have to have the best . . . regardless of sex, race, whatever. To tell you the truth, I never much thought that I couldn't do anything."

Money, who was described in the news reports as being very quiet during the conference, announced that he would be leaving the astronaut program at the end of the mission. As it turned out, Bondar, too, decided to leave the program after the flight, but it took her several months to reach that decision as she tried to assess what role she might have in the Canadian astronaut program over the following years. She was not considered by CSA officials as a candidate for the mission specialist training program that began in the summer of 1992 — a fact that she found annoying — and she was not optimistic about the near-term prospects for another life sciences flight (see Chapter Four).

If she were not training for another flight, she decided that her major career objective would be to use the knowledge she'd gained in her first flight in a concentrated research program aimed at learning more about how the human body adapts to microgravity. She pointed out that she's the only neurologist who has flown a space mission, and she believed she could put her experience to the best use by establishing a base at a Canadian university and working with NASA and the Canadian Space Agency on projects focussing on long-term space flight. If she had stayed in the astronaut program, she felt her time would be consumed with administrative and committee work and participating in a variety of short-term research projects developed by other scientists. In short, she believed she could do a lot more for the space program by getting out of the astronaut program than by staying in it.

"As a physician, I feel I have a lot of insight into medical problems and medical operations, and I really did not feel that the middle management of the agency understood, or still understands, what that means to Canada. I did not see the agency giving me the opportunity to make use of this experience." In her opinion, CSA was giving disproportionate attention to the engineering and technological aspects of the program and not concentrating enough on the human element. "The main problem with space flight in my view is not the lack of ability to develop the technology to go to the moon again, or to Mars, but rather the understanding of what physiological changes are occurring in humans. People just do not consider that we've got very low-tech knowledge of how the body is reacting to space flight. Space flight is not *Star Trek* stuff — that's very glamorous, but that's not the reality. The reality is that your body is having to cope with an environment that is very foreign to it."

4

Mission Specialists

A Canadian should walk in space . . . but who will do it?
— STEVE MACLEAN

For more than a quarter of a century, television has brought us images of white-suited, gold-visored astronauts floating against the black backdrop of open space. The stark contrast between the emptiness out there and the life support system known as earth below emphasizes the vulnerability of the astronaut and underscores the risks faced by those who are literally taking humanity's first steps out into the universe. Space walking — or, as the NASA jargon mill would have it, extravehicular activity (EVA) — is an exhilarating adventure, but one of the most demanding and dangerous. Until recently, it was something that Canadian astronauts could only dream about.

Then, in 1990, NASA invited them to join the club, offering for the first time to accept two Canadians into the training program for shuttle mission specialists. This was a particularly important opportunity because NASA had become increasingly restrictive about flying payload specialists after the *Challenger* accident. Even before that, payload specialists did not fly as often as mission specialists — in fact, many flew only once — and they were not generally perceived within the NASA organization to be career astronauts. One thing that contributed to this perception in the past was the fact that the category embraced both scientists and engineers who were doing experiments on the shuttle and others, like a Saudi Arabian prince and U.S. senator Jake Garn, whose presence was clearly little more than a PR exercise. Not surprisingly, many U.S. astronauts, particularly those who'd waited years to get a flight, found it irksome that scarce crew slots were being given to such people. But it was the death of teacher Christa MacAuliffe in the *Challenger* accident that precipitated a major change in NASA's policy of allowing "civilians" on the shuttle. Subsequently, the criteria for payload specialists were tightened up considerably; NASA declared that henceforth it would require much more justification for flying payload specialists as opposed to assigning its own mission specialists to do on-board work associated with shuttle experiments.

Roberta Bondar and Steve MacLean were both payload specialists, but NASA had made a commitment to fly them before the new post-*Challenger* policy was established. By 1992, most of those prior commitments had been honoured, and Bjarni Tryggvason commented that "as far as I know, nobody has tested the payload specialist policy that has come out of NASA since then, so we really have no idea how difficult it's going to be to fly more payload specialists."

Mac Evans, CSA's vice president of operations, said, "The conditions appear to be the following for a payload specialist to fly: One is that NASA has to have an interest in the experiment, [and] secondly, you have to demonstrate that none of the mission specialists can do the job."

Parvez Kumar, the training officer for the Canadian astronaut program, believes that in future, NASA's first instinct when presented with a request to fly a payload specialist will be to say, "'Our guy can do that. Why do we need your person?' That's why our experiments have to be absolutely unique, developed with a particular astronaut going along the whole way, so that only he or she can actually handle the experiment. That's the only way we can justify putting them on board as payload specialists. NASA holds all the cards, but we can force their hand in ways by coming up with unique things like the space vision system. If they are interested in the same thing, then that opportunity comes much quicker."

But it's unlikely there will be a large number of such unique experiments, so the tightened restrictions on flying payload specialists had serious implications for the aspirations of the Canadian astronauts, who certainly viewed themselves as career astronauts no matter what tag NASA put on them. "We're called payload specialists but in the true sense of the word that's not what we are," said Steve MacLean. They had all made a serious commitment to the astronaut program — a commitment that, as events transpired, profoundly altered the course of their professional lives. So it was not surprising that they wanted to be thought of as more than mere passengers.

This attitude appeared to have changed to some extent by the time MacLean and Tryggvason went to Houston in the summer of 1992. They certainly felt they were able to establish an excellent working relationship with their NASA crewmates and trainers. Perhaps the NASA astronauts had been mollified by the increasing flight opportunities. "Everybody gets lots of flights now; they get flights every couple of years," said Ken Money. He added, "I don't think the American astronauts would consider members of the Canadian astronaut program as being the same as they are, but it is not a big deal. They certainly do accept our guys as payload specialists, and they realize that the foreign payload specialists are part of the scene. I don't think we are looked down on or resented or . . . thought of as not real astronauts. Once you get working on a mission, you're working with them every day for two years. You do a lot of travelling together, you eat together, jog together, go swimming together, go to movies together. There's complete and utter acceptance between individuals."

Nevertheless, it seemed likely that Canadian astronauts would have a much easier time getting shuttle flights if they could qualify as mission specialists. Even more important, mission specialist skills would be essential for astronauts hoping to serve aboard space station. Station crews will be divided into three categories: station operators, station scientists, and payload scientists. The first group will be primarily responsible for station operations and for controlling, maintaining, and repairing its systems. These astronauts will most likely be responsible for operating the station's manipulator arms.

Secondarily, they will also support scientific and technological payloads. Station scientists will have many of the same skills, but their primary responsibility will be operating the scientific experiments and payloads. Both groups will be trained for EVA. Together, these two groups will perform duties similar to those done by the shuttle's mission specialists. The third group, payload scientists, will be more specialized and will fly only when a payload requires someone with unique scientific or technical skills that cannot be provided by the other astronauts. Their role will be similar to that of a payload specialist on the shuttle.

Because Canada is making a billion-dollar investment in the station and will also be contributing to its running costs, it is entitled to share in controlling station operations and to have its astronauts serve as station operators and station scientists. And, since Canada cannot afford to build its own laboratory modules for the station, its contribution of the mobile servicing system was intended also to buy Canadian scientists and engineers access to the modules built by other partners.

Canada wants most of its astronauts to be station operators or station scientists, according to Mac Evans. Although Canadians will also serve as payload scientists, it's unlikely they'd be allowed to take up all of Canada's 3 percent allotment of crew time. Evans said this would require making repeated claims that "there's a unique scientific skill that only Canada can provide" — a tough sell. "We believe the only way we can use our 3 percent is if we're in the mission specialist category of career astronauts. Rather than being one-shot specialists, we're trying to broaden the base. It's the only way that would allow us to maximize our utilization of the station and get the expected benefits from the program."

As for the astronauts, they'd been lobbying hard for mission specialist training, feeling that it represented, in Marc Garneau's words, the logical "next phase in the development of the Canadian astronaut program." Steve MacLean commented that, in reality, it would be "station operations training. We have to have mission specialist training because our hardware participation on station is the mobile servicing system. That is a flight operational thing — putting the station together, making it work, and fixing it. We have to have astronauts who can handle it, and the only way is to get mission specialist training. I think this is crucial to our success in the future."

Bjarni Tryggvason added that these skills will also be needed to help Canadian scientists develop experiments for space station. "It's not just because we're selfish and want to have space flights," he said (although he added as an aside in the next breath that "we are, of course, and do"). "If we want to give good support to Canadian scientists, we have to have people in Canada who are familiar with how you work in space and how you design for space. When you start looking at what it's going to cost to do experiments on space station, it's essential that we have people who have some experience. You will simply not be able to rely on U.S. astronauts because they just don't have the time; they are busy doing their activities."

As a first step toward this goal, CSA, in 1989, asked NASA if it could provide specialized training for Canadian astronauts in EVA procedures and remote manipulator system (RMS) operations. The main objective at that point was not to have them qualify as mission specialists for shuttle flights, but to enhance their ability to provide support and advice to the engineers developing the mobile servicing system. "We had hoped they would come back and be able to contribute to the [manipulator] designs and participate in design reviews, tests, and evaluations with a somewhat greater experience," said Bruce Aikenhead. "We could have had people with the benefits of at least some of the more formal RMS training and some measure of EVA training, so they could give opinions with a greater authority on the general problems that an astronaut might have in attempting to do EVA work or RMS operations."

During the development of the shuttle's manipulator arm, this role had been filled by U.S. astronauts. In part, this was because the Canadarm, despite its name, really belonged to NASA. (Canada paid for and gave the first arm to NASA, which subsequently bought three more.) "It was basically an American show," said Marc Garneau. "Because it was a U.S. arm, there was a whole slew of American astronauts who came up [to Canada] and provided the operator input in the early days of testing and figuring out exactly how they were going to build the controls for it."

But he pointed out that it's a different situation with the station's manipulator system, which will remain Canadian property. "It's not NASA's equipment, it's ours. The philosophy is different here; we own a piece of the space station. We would prefer it if the same process wasn't used this time, in the sense of bringing up American astronauts and asking them how the system should be operated. We'd like to have Canadians involved as well. But Canadians are not going to be very useful unless they have more than the payload specialist training, which doesn't provide any systems training in RMS or any other things like that. If we could get Canadians with mission specialist training, with a specialization either in RMS or EVA, then they could provide a greater Canadian contribution to the operational development of the MSS and there would indeed be a Canadian presence."

NASA turned down CSA's request for special training for the Canadian astronauts. In a letter to Evans, William Lenoir, then NASA's associate administrator for space flight, said their training teams and facilities were already overloaded supporting upcoming shuttle flights and the space station program, and "adding to this work load by providing the requested training will not be possible." In any event, he said, the limited experience it would provide the Canadian astronauts wouldn't really give them an in-depth appreciation of space station operational problems, and he offered to make seasoned NASA astronauts available to CSA to assist in design reviews and development testing of the MSS.

But NASA was not unresponsive to CSA's objectives: Lenoir instead offered to accept two Canadian astronauts for mission specialist training at the Johnson Space Center starting in the summer of 1992 — an offer that, in the long run, would probably prove to be more beneficial to the Canadian program than what CSA had been asking for in the first place. Anticipating the time when international crews would routinely be working together on the space station, NASA had made this offer to all the international partners and was expecting to take in European and Japanese mission specialist candidates as well.

This was an unprecedented opportunity, but it meant losing the day-to-day services of two astronauts for up to five years. While the mission specialist program would provide the Canadian astronauts with the desired EVA and manipulator training, the deal was that they'd be seconded to NASA's astronaut office, where they'd become involved in many other activities under NASA's direction. These so-called "collateral duties" included such things as working in Mission Control, participating in safety reviews of shuttle equipment and experiments, and advising and supporting companies, universities, and other organizations flying space experiments. "From time to time, they are assigned to work with some particular department within NASA, learning what's involved in safety, or life support systems, or remote manipulators, or what have you," said Aikenhead. "They participate in the day-to-day work, especially when designs are being reviewed, presenting the astronaut point of view and also bringing back to the astronaut office news of what's going on in that particular area. Collateral duties also include serving on committees and participating in special studies.

These activities would provide the Canadian astronauts with important insider experience in space flight operations. But, more significantly, being part of NASA's mission specialist pool would give them a shot at extra shuttle flights. In his letter, Lenoir said NASA intended to offer a flight to "one or both" of the Canadians, who would undergo a year of formal training followed by on-the-job training within the NASA system until being assigned to a flight. Lenoir noted, however, that it usually takes four to five years from the start of training to receiving a flight assignment, and he said the Canadian mission specialists could, if CSA preferred, "return to Canada to support CSA's space program, if you elect not to wait for a space shuttle assignment. However, this would miss the main benefit [of mission specialist training] — space flight operations experience."

In December 1990, Evans responded by formally accepting NASA's invitation. He confirmed the Canadian astronauts would spend the full four to five years working at NASA, including the on-the-job training and collateral duties. "In accepting your offer, we . . . envision an arrangement whereby two Canadian astronauts are absorbed into the NASA corps of mission specialists awaiting mission assignments, and they would be eligible for any duties or assignments open to any of the other members of the corps." In an interview, Evans noted that this meant Canadians "could be chosen for a flight that has absolutely nothing to do with Canada or they could be chosen for a flight that has something to do with Canada. It's open." (Another CSA official, however, expressed some doubts on this score. Asked if he thought NASA would select a Canadian for one of their own flights, he said, "Probably not. Would you select somebody from the United States on your flight? I wouldn't. I think everybody is very nationalistic.")

Evans said the job would be equivalent to being "posted abroad for four or five years. We're talking about a fairly long-term commitment on the part of the people who go down." But CSA did not want the Canadian astronauts, as Aikenhead put it, "to go totally native down there, so from time to time they'll need to come back home." Bob Thirsk commented that even though the Canadian mission specialists would be busy with NASA activities while awaiting flight assignment, "we would hope that in this time frame they would certainly be allowed to come back to Canada and support the Canadian engineers on the MSS development or some of the Canadian space scientists on the development of scientific payloads. We certainly would like to have them maintain continuity with Canadian scientists and engineers and the agency, too. We don't want them to lose too much perspective about . . . the Canadian space program." And Garneau, who had prepared a report on the benefits of mission specialist training, expressed confidence that NASA would try to accommodate Canada's interests, even though the U.S. agency had "made it clear that under this concept the Europeans, Japanese, and Canadians who become mission specialists are a NASA resource — that essentially they decide what these people are going to do."

Evans, too, felt that NASA would "recognize that they'll be

our employees and there are things we'll want them to do. NASA won't have them full-time and neither will we. We get the experience of having people who have worked in the program. That's why the collateral duty business is important — so we're able to draw on that experience from time to time." Aikenhead also foresaw long-term benefits for the Canadian astronaut program itself. "When we go into the pool of space station astronauts, we'll have acquired more people with actual flight experience that could be passed on to other members of the astronaut program."

Thus, in his letter to Lenoir, Evans took pains to emphasize CSA's expectation that an effort would be made to ensure the Canadian program could benefit from the astronauts' training and experience. "It will be both appropriate and essential for the Canadian astronauts, during their tour of duty with NASA, to retain their connections to the Canadian Space Agency in support of CSA functions, including public affairs activities. In all other respects, they would be on equal status with their NASA colleagues, and their daily schedules would be in accordance with the wishes of the NASA astronaut office. With regard to the on-the-job training, the collateral duties, and ultimately the shuttle flight assignments, we would expect that due consideration would be given CSA activities and that NASA would consult with the CSA prior to making decisions in those areas."

Evans concluded his letter by confirming that Canada would continue to cover its astronauts' salaries, benefits, accommodation, and travel expenses, in addition to paying NASA's out-of-pocket expenses for their training — roughly half a million dollars a year. (In effect, NASA is getting two new astronauts for free.)

When Lenoir responded in April 1991, he had concerns about CSA's desire for consultation on assignments given to the Canadian astronauts. "Unfortunately, due to the complexities involved with collateral duty and on-the-job training assignments, and the desire to treat the [international partners'] astronauts as much like U.S. astronauts as is reasonable, it would not be advisable for NASA to consult with CSA prior to the assignment of such duties," he noted. "However,

NASA does intend to consult with CSA prior to assigning any Canadian astronaut to a shuttle mission."

He also clarified another point: while it was NASA's intention to offer a shuttle flight to each of the Canadians, it could not guarantee that both would actually fly. This depended on the rate of shuttle flights, as well as "mission objectives and overall operational requirements." However, Lenoir reiterated his assurance that non-U.S. astronauts would be treated as much like their U.S. colleagues "as is reasonable" with respect to flight assignments. Theoretically, this meant that the Canadians could be assigned to NASA flights in which Canada played no part.

Aikenhead wrote back to Lenoir saying that CSA accepted the fact that NASA couldn't consult CSA about every collateral duty assignment, which he conceded would be "altogether an unwieldy nuisance." However, he said, "we believe that there should be a mechanism whereby the CSA could review the collateral duty assignment record of its [mission specialist] candidates, in relation to their backgrounds and possible long-term future roles in the CSA, and provide input concerning preferences for upcoming collateral duty jobs." He noted that CSA would have concerns if, for example, the Canadians were sent to work at various NASA outposts while others were given projects related to space station development.

In an interview, Aikenhead acknowledged that CSA expects a better fate for the Canadian mission specialist than to be stuck "way off in the corner. If we found that our people were getting all the dull jobs, we would understandably be upset. We would find it strange, certainly, if there were no RMS involvement." Collateral duties related to life sciences would also be "in keeping" with Canada's interests, he said.

On May 21, 1992, a formal agreement was signed between CSA and NASA. Aikenhead and Evans were then faced with the task of selecting the lucky candidates. This became rather a delicate issue around the astronaut office, because they'd decided to wait until the new astronauts were on board before making their selection. Their reasoning was that since the new recruits would be evaluated with mission specialist and space station criteria in mind, there might well be some among them who would be exceptionally well-qualified for mission specialist training. Aikenhead said they did not want to "rule out people that we didn't even know about yet. It could become perfectly obvious that there was a candidate we really ought to put forward, [and] we felt it inappropriate to say they'll come only from the existing group [of astronauts]." He denied there was an arbitrary decision to select one each from the two groups of astronauts. "We said we wanted to keep our options open on this one."

Most of the existing group were, however, keenly interested in the opportunity, although the timing was wrong for some of them. MacLean, for example, was clearly out of the running for the 1992 program because he would then be in the final stages of training for his flight. Ken Money had already informed CSA that he would be leaving the astronaut program after the IML mission. "What we were looking at was a further five years, and Ken said he was against it," Aikenhead said. As for Roberta Bondar, she would have her hands full dealing with the aftermath of her mission. "Roberta has been very heavily committed to all the other things she already does," Aikenhead said. "She was not seriously considered . . . as a mission specialist." (To which Bondar later tartly responded, "Who set the schedule?")

That left Garneau and Thirsk of the original group as the two most unencumbered by existing obligations. As for Tryggvason, Aikenhead conceded at the time that his situation was one that "we would clearly have to agonize over." As backup astronaut for MacLean's mission, he would still be in training when the 1992 mission specialist course began, and CSA officials regarded this as a serious schedule conflict. They did not believe it was possible for Tryggvason to do both jobs at once. "The only thing you could do is to say we don't have a backup or we find somebody else as a backup. That's not terribly practical," said Aikenhead. The astronaut office was short-staffed and, in any event, it would be difficult for anyone — even an experienced astronaut — to jump into training for a shuttle mission at such a late stage, particularly when the major objective is to test a complex piece of hardware like the space vision system. "It's a highly technical experiment," said Mac Evans. "It takes a lot of training."

However, Evans and Aikenhead hedged their bets a bit by asking Tryggvason to take the mission specialist medical while he was down in Houston, in case one of the other candidates failed on medical grounds. "So, in principle, I guess I was in the running, because the CSA arranged for me to have this endless medical," Tryggvason said. Aikenhead said that if one of the others hadn't made it, "we would probably have tried to develop a work-around for Bjarni, so that he might be able to get in as the mission specialist, and somehow, if Steve broke his leg before the flight, Bjarni could step in."

Tryggvason believed he could have handled both jobs and felt his role as MacLean's backup should not have been an impediment "to the extent it was. We talked about this with people here, and the NASA folks really didn't have any great problems with this. Most of our training on a lot of experiments had to be done by the end of July, and there was not a lot of real conflict. I don't think that should have been a major factor, but the people at the CSA did consider that to be a big stumbling block and held the view that they wanted to maintain the backup role rather than put me in the mission specialist role. I'm disappointed that I was not selected; I would have liked to have done that [and] I think I would fit in very well with the people down there."

CSA did not view the medical as wasted effort. "It tells us, the next time we get this opportunity, whether or not he is in the running, at least medically," Aikenhead said. But Tryggvason wasn't sure there'd be another opportunity. "This current input is a one-shot affair. There's no certainty that NASA is going to invite us to put in more mission specialists when they hire their next batch." (While this was true, CSA officials said they were confident there would be future opportunities for Canadian astronauts to become mission specialists.)

Garneau was very keen on mission specialist training, which was very much in keeping with how he wanted his career as an astronaut to evolve. "I think probably I am more of an operator-engineer type of person than a scientist," he said several months before the selection was made. "My career is going towards the more practical side of operating and keeping the systems going. I like hands-on activities. I do things that involve the mind and the body — I like to parachute, scuba dive. Certainly something like EVA brings that all together. I would love to operate the arm out in space. I've done it on the ground; I'd like to do it out there. If I were asked to become a mission specialist, I would say yes. I would be very flattered, and I would be prepared to make that commitment. I've been spoiled, already having had a mission; this would be icing on the cake."

Throughout the latter half of 1991, the unresolved question of who would be assigned to mission specialist training caused a certain amount of tension within the astronaut office. Those who were keen on the assignment were diplomatic about having to wait until the new recruits were chosen, but it was only human nature for them to feel that their years in the program, their patience in outwaiting the innumerable delays in shuttle flights, and especially their firsthand knowledge of NASA's complex training and mission-planning operations should count for something.

Garneau had given considerable thought to both sides of the question. He observed that "NASA takes people fresh off the street and turns them into mission specialists. You could use the argument, why can't they take Canadians fresh off the street and do it?" Obviously, NASA could. On the other hand, he said, "I think we would probably argue that we're higher up on the learning curve than new people coming in, and that that should be a factor taken into consideration. But in the end, we're not the ones who make that decision. Obviously, each one of us would like to be considered for it. We feel that we should be eligible."

At that time, it was uncertain whether the recruitment campaign could even be completed in time, but Aikenhead said this would not prevent Canada from accepting NASA's offer. The existing astronauts would make excellent mission specialist candidates, and "there was never any question of our turning down the invitation. It's too important to the

Canadian program — an opportunity just not to be turned down."

In the end, the timing was tight. The new astronauts were selected on June 6, and Garneau and Chris Hadfield were informed shortly afterwards that they'd been chosen — subject to passing yet another medical administered by NASA. "We had a bit of a chicken-and-egg situation," said Aikenhead. They didn't want to make a public announcement until they were sure the two would meet NASA's requirements, "so we quietly arranged to have the medicals done." This was the only respect in which NASA had any involvement in the selection; otherwise, the decision was entirely up to the CSA. Obviously, however, Aikenhead and Evans were guided by NASA's own mission specialist criteria and also by the criteria established by the international partners for space station astronauts. "We had the job of choosing people who were going to represent Canada in yet again another unusual kind of situation — they were going to be pioneers, the first Canadians to train as mission specialists. We had to be sure we were picking the best of the available people," Aikenhead said.

The difficulty was that all the potential candidates had many excellent qualities and skills — they wouldn't have made it into the astronaut program without them — but they each had different strengths, so it became a matter of juggling disparate factors. One overriding factor, in Hadfield's case, was that he had superb test-piloting skills and had acquired most of his training and experience in the United States, which meant he would likely fit into the NASA culture quite easily. "He had the kind of experience that would gain him instant acceptance as a mission specialist," said Aikenhead. "He has already worked with a bunch of the guys down there over the years." As a test pilot, he had also demonstrated the intelligence and the ability to cope with stress needed of a mission specialist. As for the other three new astronauts, while they all had impressive credentials, "we felt they were not quite there yet. We needed a bit more time."

CSA's decision to send two astronauts with engineering backgrounds was later criticized by Roberta Bondar after she resigned from the program. Although she certainly would have liked to have been considered for the position, she emphasized that her objections stemmed not from the fact that she wasn't chosen, but that it was shortsighted not to consider sending an astronaut from a nonengineering discipline. "I'm not trying to be a cry baby and say they didn't choose me so I'm quitting. It's the principle of putting all their eggs in one basket when we're looking towards space station. I think they could have covered their bases more. We need to have people who are trained in science, who have a mind to do science, to be on board space station. There is no scientist down there training as a mission specialist."

She pointed out that NASA trains people from all kinds of backgrounds as mission specialists, and that Canada's choice not to do the same was like "saying Canadian MDs or Canadian geologists or Canadian whatevers are not as good as American whatevers. They're not giving the nonengineers any credit that they can do this task, when [NASA's] already giving that credit to American nonengineers." She also questioned the decision to send for advanced training someone who had not yet completed his first year of probation as an astronaut candidate.

Mac Evans's response was that there was a known, near-term requirement for Canadian astronauts to fly engineering flights — notably the shuttle mission on which the mobile servicing system would be checked out — and "we were simply looking at what was on the deck as a certainty. We were not downplaying the life sciences at all. There will be more mission specialists going down. We hope all of these people will get their mission specialist training and there will be, I hope, an opportunity for a life scientist." He acknowledged that there was, at that point, no formal agreement or firm commitment from NASA to train more Canadian mission specialists, but "in informal talks with them, it was made quite clear this would be an ongoing thing. I just cannot see how they can pursue the international space station without having everyone go through this eventually." And, while CSA intends to pursue this, "we're letting our two guys down there now pave the way and make sure the field is fertile."

As for Chris Hadfield's still being on probation, Evans said there was a risk that anyone who was sent to the mission specialist course might not make it through, "but we felt that the risk was very minimal, given his background."

In June 1992, Hadfield and Garneau went tearing down to Houston to be poked and prodded by more doctors. For Hadfield, the experience was just another in the long line of seemingly endless tests that had come his way over the previous six months. Taking the psychological test, he discovered that the Canadian test that he'd found so long and tedious had actually been a short version, containing only "the greatest hits" of the NASA test. "With NASA, I got to do the entire 600-question exam. They also did IQ tests, and much more in-depth. It took forever; it was four hours of written exams and then a three-hour time with the psychiatrist."

For Garneau, who combined the four days of examinations with SVS business he'd already scheduled in Houston, it was his first encounter with the exhaustive mission specialist tests. "A couple of weeks later when all the results were in, I was told that I had passed and to stand by. Towards the end of June, they told me I'd been selected . . . just very quietly, no big deal. Bruce and Mac notified me that I had been formally selected by them." But he wasn't letting his euphoria blind him to the hard slogging that faced him. "Boot camp lies ahead of me," he said. Even though he'd learned a lot about the NASA system over the years, first during training for his own shuttle flight and then doing administrative work on the SVS mission, he was not going in with the attitude that "I've been through some of this before, and therefore I'm going to be able to coast through it. I've got to go in there and work my tail off."

Asked how he thought the Canadian public would react to the idea of Canada's first astronaut going to work in the United States, he suggested that "it wouldn't be a shock because I think the majority assume that's the case anyway. There are a lot of Canadians who still think that Bob [Thirsk] and I worked for NASA, and they're surprised that we're in Canada. They don't see a contradiction in your being a Canadian and yet working for NASA because they're just not aware that you actually work and live in Canada and support Canadian scientists. We still do a lot of correcting of impressions on that score." He added that if the Canadian mission specialists were, as hoped, involved in supporting the space station MSS program, they "could be making frequent trips to Canada as part of that, because that is where the hardware is and the work will be done."

Near the middle of July, a few days before he left for Houston, he commented that "it's a very heady time for me." And not just because of the new job, he added. "I'm getting married tomorrow, so there are lots of things happening in my life. I wish it had all been worked out earlier, but that's too much to ask. I'm just scrambling. I don't plan it that way ..." But he had no qualms about moving to Houston; in fact, he'd been spending half his time there anyway, mothering the SVS mission through NASA's paperwork labyrinth. "I know the Houston area well, so it's not a strange place. I've been down dozens of times, [and] I know people there."

His new wife, Pamela Soame, had been a nurse with the Canadian Forces for several years and was then completing a nursing degree at an Ottawa university. Garneau had talked over the potential move to Houston with her and said she was "looking forward to a change. Getting married is quite a big change, so changing towns is not much of an additional adjustment. She's got one more year of nursing at the university to complete, and she'll probably transfer to the university in Houston." As for his twins, although they had mixed feelings about leaving their friends, "they've been down there, so they know the place as well."

Garneau and Hadfield reported to the Johnson Space Center in early August 1992 to begin basic training — 28 weeks of lectures, briefings, tours of NASA facilities, and operational activities. The core of the program consists of a thorough review of the space shuttle and its operations. The astronauts learn about the basic theory, design, and operation of all of the shuttle's major component systems: data processing, electrical power, propulsion and manoeuvring, mechanical, environmental control and life support, navigation, communications, and the camera/TV systems.

Once the basics have been covered in a series of briefings and lectures, the astronauts will be given "hands-on" training in NASA's many simulators and training facilities, including an earth-based version of the Canadarm and the Weightless Environment Training Facility (WETF). The WETF is a large pool in which astronauts clad in bulky white space suits learn EVA procedures on full-scale mockups of the shuttle's cargo bay and the payloads it carries. The slow-motion movements under water are a good approximation of the kinds of movements astronauts make in microgravity. To qualify for working in the WETF, mission specialist trainees also take scuba lessons.

Other components of the training program include lectures on life sciences, basic electronics and computers, and applied science subjects, including space physics, geology, planetary science, atmospheric sciences, astronomy, star identification, earth visual observations, oceanography, and materials processing. These lectures are intended to give the trainees the basic knowledge of scientific and technical subjects that they may one day need on a shuttle flight; one of the major duties of mission specialists is to supervise scientific experiments and to monitor and fix equipment on behalf of earth-based scientists who have spent many years and millions of dollars preparing experiments to fly on the shuttle.

Finally, since astronauts spend a lot of time flying in NASA's small T-38 jets, the mission specialist trainees will also receive T-38 briefings and parachute training and will be taught both land- and water-survival techniques. "That prepares them if they have to eject out of T-38s over water or land, or even in the shuttle, where they have to get out in an emergency situation. It teaches them getting out of parachutes in water, those type of things," said Travis Bryce, a NASA official who served as the mission specialist liaison with the Canadian Space Agency.

After about nine months — presuming they pass basic training — the mission specialists will move on to advanced training designed to hone their skills further and maintain their proficiency in what they've already learned. And they'll be assigned to collateral duties. "That's where we actually assign them technical tasks within the astronaut office, and they will have a specified job to do as part of our routine operations on a day-to-day basis here," said Bryce.

"You could be assigned to be the specialist on safety reviews, or you could be employed as a Capcom for a mission," said Garneau. (A Capcom is an astronaut who communicates directly with the crew in space.) "You could be doing things that astronauts do in support positions, but may not have anything to do with Canada's involvement in space station. We've said it would be nice if Canadian mission specialists could be employed in fields related to the development of the space station, because some of the American astronauts will be assigned to space station development. And particularly it would be nice if at least one of the Canadians could be specifically assigned to the MSS operation side of things."

"We would like to treat each mission specialist the same as much as possible, and we like to give them as much cross-training as we can," Bryce commented. "We want to make any one of them capable of doing any job on the shuttle." He said that the mission specialists do not spend a great deal of time working on any particular collateral duty. "They're assigned for a short period of time, and then we rotate them into another job. The time in the job depends on what our requirements are and how many people we have on board. We try to take the best advantage of the background that each has when they come on board . . . give them jobs where they can pick up without a loss of continuity and effectiveness. We're going to assign them . . . where they can best benefit the shuttle program. We'd like to benefit both agencies, but these are going to be mission specialists that we're going to train to operate the shuttle. Part of that is operating the arm and the robotic gear, but that's only part of it. We're going to train them to do anything on board the shuttle that anybody else can do."

One of the major ways of doing this is running the astronauts through intense and demanding exercises in computer simulators, starting with the simplest ones, known as "single system trainers." Bryce said these allow the astronaut to "let their hands work with their brains to try and figure out how systems work and what the displays on board are. That's very basic — we teach each system to the extent they need to know it. When we get past that point, we move to a more integrated or complex type of training situation, where we put the whole thing together. When we take one further step beyond that, we try to combine it all into the shuttle as an entire vehicle. That's when we go over to our mission simulator . . . where we try to train the mission specialists and the pilots to handle any failure that may arise in any system, and oftentimes there are multiple failures. The system's so intertwined that unless you understand how the systems operate, one failure may mask another. As you start doing these things, we get them familiar with our procedures and malfunction checks. They have to isolate where the failure is and what the corrective actions are."

By the time they do these exercises, they're getting beyond basic training. "There's a transition period between nine and eleven months, and then they start getting into very serious advanced training," said Bryce. It's at this stage that they'd start to get EVA and RMS training. Aikenhead said all mission specialists receive EVA training because NASA has a policy that each shuttle flight will have at least two crew members qualified to do space walks. Although EVAs are usually carefully planned and rehearsed before the mission, there have been occasions when astronauts have done unscheduled space walks. "It's generally in their interests that all of the mission specialists have had that training."

By the third or fourth year, mission specialists should be ready for flight assignment. Bryce said that after the first two years, "it becomes essentially a competition for crew assignment. The people who proceed through their training well and blossom would probably be assigned." He added that, as far as the astronaut office at the Johnson Space Center is concerned, the Canadian, European, and Japanese mission specialists would be eligible for assignment to NASA flights. "I know headquarters has a different perspective. They may surmise that, for example, the ESA people are more closely tied with ESA. We would like to hold the option that if there's a flight available and, for example, the Canadian person is the best person to fly and does not have Canadian payload, we would like to hold the option open to fly on that flight."

He also said that NASA astronauts are "without a doubt"

willing to accept the foreign mission specialists. In fact, "we need some new astronaut bodies here." A number of NASA astronauts had recently left the program, and "right now we need to bring new people in to fill the gaps."

After a flight is assigned, the mission specialist would then undergo at least another year of even more intense and narrowly focussed training. "Once a mission crew has been identified, they start training towards that particular mission quite specifically," said Aikenhead. "If it's a mission that requires EVA — let's say it's a satellite retrieval — then they'll have to become very proficient in exactly what tools to use and how."

The MSS verification flight scheduled for 1996 will require the services of mission specialists, both to operate the arm and to do a space walk to unlatch the folded boom arms of the large manipulator arm. CSA officials expect the Canadian mission specialist to perform at least one of these tasks. "I'm not sure what else you could do [that would be] associated with checking out the MSS," said Mac Evans, who made it clear Canada expects far more than a sightseeing role for its astronaut. "We're supposed to know our equipment better than anybody else, so we should be the ones to check it out."

When Karl Doetsch, director of the Canadian space station program, was asked whether he thought Canadians will soon be seeing one of their own floating free in space outside the shuttle, he said emphatically, "I'm expecting it."

5

On the Flight Deck

Every minute is a lot of fun. It can only be like that a few times in your life, and this is one of them.

— STEVE MACLEAN

They made Steve MacLean and Bjarni Tryggvason jump out of the shuttle. They bounced them around in a centrifuge. They made them put out oil fires. And they threw them into the water over and over again.

But the two astronauts insisted they were having a great time.

"It's like you're at summer camp," said an exuberant MacLean about three months before his scheduled nine-day shuttle flight in October 1992. He and Tryggvason were then about half-way through the intensive final training program at the Johnson Space Center in Houston. "We did three days of water survival out in the ocean, and that was great — got a lot of beer stories from that one. Another special session was the centrifuge — you go to Brooks Air Force Base in San Antonio and ride the ascent profile." The ascent profile simulates the gravity forces an astronaut experiences during the shuttle's fiery climb through the earth's atmosphere. The shuttle's lift-off generates forces three times normal earth gravity, which MacLean describes as "pretty docile," so the simulation "was not like a ride on Astro World, but it was still very interesting." (No doubt it would have been even more interesting had the machine functioned as described by the Canadian Press. Beneath a photo of a suited MacLean sitting in the centrifuge chair, the caption stated that the device "simulates 30 times the force of gravity that spacemen experience during liftoff.") After experiencing the real thing, however, MacLean acknowledged that the launch was a "kick in the pants."

"We went through fire-suppression training," added Tryggvason. "We were putting out oil fires and gas fires, and that was fun, too."

Of course, these activities served a serious purpose: they were designed to teach MacLean and Tryggvason (should he have been required to step in at the last minute) all they needed to know to function efficiently and safely as part of a shuttle crew in both normal and emergency situations. Like Bondar and Money, they received more extensive safety training than that given to Marc Garneau before his flight — another legacy of the *Challenger* accident. (Garneau spent two months before his flight in Houston, compared with about six months for the others.) In MacLean's case, an added factor was that he would be spending a great deal of time on the flight deck, the command centre of the shuttle, where mistakes could potentially have very serious consequences.

One of the things MacLean particularly appreciated was that NASA allowed his wife, Nadine Wielgopolski, to watch many of the safety training exercises. "They make it easy," he said. "When they dunk us in the water or do a bailout out of the orbiter, I try to make sure she's there." In fact, about a month before the flight, she participated in an exercise herself. "She gets to ride the ascent simulator. It's part of our regular simulation; she sits beside me and gets to feel what it's like. I think it's a very intelligent thing to do. It gets her closer to what I'm feeling like during the ride, and it also gives her a tremendous amount of confidence in the system." MacLean said this practice had been started in response to a recommendation by NASA astronauts, who felt being more involved would help their families handle the demands of their job. "The sacrifice the family makes is quite high; anytime you get somebody more involved, they understand it better and you get more support. It's very well received."

He was especially glad his family could be with him during the final six months. His wife was on maternity leave from her job as head of official languages at the National Research Council, having given birth to their second child, a girl, in April, shortly after he'd moved to the Johnson Space Center. (The couple met while Wielgopolski was giving French lessons to Roberta Bondar.) "If she was up in Ottawa, we wouldn't be able to talk each day," MacLean said. "I feel great that she's going to know a fair amount of detail about what I'm doing and what it takes to do it. It's going to be a really neat thing to share; I feel pretty lucky."

It also allowed him to spend time with his children, something MacLean greatly valued but was finding increasingly difficult to manage as his launch date approached. "Family life ... forces you to be a little more disciplined in your schedule. It forces me not to do anything between 4:30 and 8:30 in the evening — and actually that's great; I have such a good time. You come home and you can escape into their world for a couple of hours." But he acknowledged that trying to help his wife with a six-week-old baby while in the final stages of preparing for a shuttle mission was no easy task and that she was having to carry most of the load. "With a new child, you don't sleep much. So for both of us, it's been a little hard."

One of the major purposes of the final six months of training is to meld the crew into an efficient, well-honed team. Therefore, in addition to the safety training, MacLean and Tryggvason participated in simulations of the entire mission with their NASA crewmates, commander James Wetherbee, pilot Michael Baker, and mission specialists William Shepherd, Tamara Jernigan, and Charles Lacy Veach. A number of these sessions took place in two facilities known as the shuttle mission simulators, the most sophisticated of which is fitted with a hydraulic system that shakes and tilts the crew compartment to mimic the vibrations and movements of the shuttle and is said to be an extremely accurate reflection of the real thing.

During these "integrated" simulations, which towards the end of the training period also involve all the support people in mission control, the crew rehearses everything that's supposed to occur during their flight — and a lot of things that aren't supposed to occur. "They're throwing malfunctions at us, and the crew has to sort those out," said Tryggvason. "These guys do not go five minutes without something going wrong. Everything has to be responded to like it is real."

"It gets more intense as we get closer [to the flight] because we become the prime crew on the simulators," said MacLean. "We've been working 12- to 13-hour days. But we're in good shape; it's a matter of fine-tuning, so that you peak for the flight and not the week before or the week after."

Tryggvason was surprised that he and MacLean were included in integrated simulations that had little to do with their own experiments. These were "generic . . . simulations for operating the orbiter — training sessions for the rest of the crew. [This] generally would not have happened, say, with Marc and Bob when they went through this. We get a much better appreciation for what's going on. In fact, they allowed us to operate some of the equipment, so that really makes you feel much more part of the crew. The stuff that we're operating is relatively low-level compared to what the rest of the crew is operating — we just do not have the same background — but it's nice to be accepted as part of the crew."

To some extent, this reflected the personal approach of their NASA crew, but Tryggvason believed there was more to it than that. "I think it comes in part from the *Challenger* accident and the acceptance by NASA that they have to prepare everyone much more. All of the payload specialists — ourselves, the Japanese, the Italian ones, and the German ones — are spending much more time down here and getting much more training."

In addition, Tryggvason said there seems to be a growing "acceptance of the fact that more attention has to be paid to the use of the shuttle for achieving good science. Maybe it's sort of the writing on the wall for the space station — that much more emphasis has to be placed on making sure that good science is achieved both on the shuttle and on the space station." Coupled with this was what appeared to be an improvement in the attitude toward payload specialists — "a general drift toward better acceptance of payload specialists as authentic astronauts. Certainly Steve and I have found that the crew has accepted us as experts on our science. They're really quite keen on some of the experiments and helping us out as much as they can."

This was particularly critical to the success of the space vision system (SVS) experiment, which was far more integrated with shuttle operations than any previous Canadian experiment. It required close coordination not only with the mission specialists operating the arm, but with the commander and pilot who are responsible for manoeuvring the shuttle itself. And, unlike the other Canadian experiments, which were carried in the lower compartment known as the mid-deck, the SVS control and display equipment operated by MacLean was installed on the flight deck, near the flight controls and the navigation equipment for the shuttle itself. It was no small accomplishment for the SVS to be given precious flight deck space or for MacLean to be accorded the right to work there, which demonstrated the extent of NASA's interest in the new technology.

"I do not feel like a payload specialist at all," MacLean commented. "We have a very complex flight technically, and I think the crew appreciates that. It's not a PR trip."

MacLean was responsible for a group of scientific experiments collectively known as CANEX-2 (Canadian Experiments-2). Testing the SVS was the major objective of his mission, but there were six other experiments requiring his attention as well. They included:

Material Exposure in Low Earth Orbit (MELEO). Specimens of several plastics, composite materials, and lubricants used in the construction of space structures, such as satellites, the space station, and the mobile servicing system, were attached to plates on the Canadarm and exposed to the harsh space environment for a total of 30 hours. Several experimental materials that might be useful as protective coatings were also tested. The degree to which these materials were eroded by atomic oxygen — likened by one scientist to burning them with a blowtorch — was analyzed after the samples were returned to earth. This erosion causes a loss of mass, strength, stiffness, and stability of size and shape. In addition, MacLean controlled and monitored 350 elements called Quartz Crystal Microbalances attached to the end of the arm, which provided him with continuous, highly accurate measurements of the amount of erosion occurring.

This experiment continues previous research on the destructive effects of atomic oxygen in low earth orbit that began after scientists discovered a shocking amount of damage on satellites and space vehicles that have returned to earth. One major concern is the potential impact on the space station and its components, which are supposed to function for at least 30 years in space. As a result, researchers have been trying to document how well various existing types of materials withstand atomic oxygen and to develop new materials and coatings that can shield them from the assault. These materials might also one day have important applications on earth, in electronics, packaging, and insulation. The principal investigator on this experiment was David Zimcik of the Canadian Space Agency.

Before the mission, MacLean said he was "entirely comfortable with this one because this was a common conversation at the supper table." His father, Paul McLean (their last names are spelled differently because of a clerical error on the elder McLean's birth certificate), was for many years a researcher with the National Research Council of Canada, and he developed materials that were among those tested on Marc Garneau's flight. McLean also developed some of the materials flown on his son's flight. It marked the first time a father-and-son team had worked together on a shuttle experiment.

*O*rbiter Glow-2 (OGLOW-2). A glow appears on the surfaces of the shuttle facing the vehicle's direction of motion, which is believed to be caused by the impact of high-speed atoms and the effect of the shuttle's surface temperature. The experiment, which involves photographing the glowing surfaces, was first carried out by Marc Garneau on his flight. The results of that experiment were surprising in that they indicated the glow occurs mostly around the shuttle's tail section, not all over the vehicle as had been thought, and that the firing of the shuttle's thrusters appears to have an effect as well. On his flight, MacLean used new, specially designed equipment to photograph the glow on the Canadian target assembly, a metal plate used for the SVS tests, especially when the thrusters are being fired. Video recordings were also made using the shuttle's on-board cameras. The principal investigator for the OGLOW experiment was Ted Llewellyn of the University of Saskatchewan.

Queen's University Experiment in Liquid Metal Diffusion (QUELD). This was an experiment to study the mixing of different metals in the liquid state in weightlessness. More uniform mixing is possible in space because of the absence of gravity effects on the movement of atoms (diffusion) that occur on earth. Researchers believe that materials-processing experiments such as this one will help them to improve similar processes for making alloys and crystals on earth, and that one day it might be economical to produce extremely pure materials in space, including some that cannot be made in earth's gravity at all. In this experiment, MacLean was required to put about 40 capsules, each containing two metals, into a small furnace, where the metals were melted and allowed to mix. Then they were rapidly cooled to solidify the metals, which were stored to be analyzed on earth. The principal investigator for this experiment was Reginald Smith of Queen's University.

Phase Partitioning in Liquids (PARLIQ). This was another materials-processing experiment. Its objective was to use microgravity to separate a liquid mixture into two very pure components or "phases." When a collection of cells is added to the liquid, some are attracted to one phase and others to the other phase, and this affinity can be used to separate the cells, too. On earth, this process can be affected by gravity, producing less pure separations. Ultimately, these experiments may lead to the development of ultrapure medicines, including ones needed for transplants and the treatment of certain diseases. The principal investigator, University of British Columbia professor Donald Brooks, hopes to use the technique to separate leukemia cells from healthy blood cells in bone marrow, something that is impossible to do cleanly on earth. (Brooks's own healthy red and white blood cells were used for the test on MacLean's mission.)

Sun Photospectrometer Earth Atmosphere Measurement-2 (SPEAM-2). In this experiment, MacLean used two instruments developed by the Atmospheric Environment Service of Canada to take measurements of the earth's stratosphere. The objective was to gain a better understanding of stratospheric chemical processes and their effects on the ozone layer, which protects life on earth from the sun's damaging ultraviolet radiation. (In recent years, scientists have discovered that huge "holes" in the ozone layer materialize at certain times of the year, believed to be caused primarily by reactions involving by-products of industrial chemicals released at ground level.) The measurements made by MacLean, which were a continuation of work begun on Garneau's flight, were expected not only to provide valuable new data on stratospheric chemistry, but also to test the capabilities of the new instruments, data that will help Canadian scientists make better use of space platforms such as satellites and the space station for atmospheric monitoring in the future. David Wardle and Tom McElroy of Environment Canada were in charge of the SPEAM experiment.

MacLean said the data would be shared with researchers in several other countries who are taking ground-based measurements. "It's all part of the world meteorological network. What's very interesting is that the improvements made in the photospectrometer have gotten an international reputation." During his preflight training, he participated in an exercise in which this instrument and those from several other countries were tested and compared on top of a mountain in British Columbia.

Space Adaptation Tests and Observations (SATO). These experiments continued the research done on Garneau's and Bondar's mission to study the effects of microgravity on the human body. Included were tests of the balancing system and illusions of movement caused by weightlessness, as well as a repeat of the back-pain experiment done on Bondar's flight. (During the flight, MacLean said his body lengthened from 178 cm to 183 cm, and he did experience some back pain but not enough to "keep me from sleeping.") In addition, there was an experiment to study the upward shifting of body fluids in microgravity that results in what astronauts call "puffy face" syndrome. This shifting also causes a loss of body water over a period of days; early results from Bondar's flight indicated astronauts may suffer significant dehydration. At the beginning and end of his flight, MacLean drank a glass of heavy water, which, because it contains oxygen with three atoms instead of the normal two, can be traced. Saliva samples collected during the flight were analyzed afterwards to provide a record of water loss. This information will be useful in developing nutritional requirements for long-duration spaceflights. Several scientists, lead by principal investigator Alan Mortimer, developed these experiments. As for his role, MacLean said, "I am just . . . a victim."

MacLean spent a lot of time on the ground before his mission practising the procedures and techniques for doing these experiments and learning how to use the equipment. For example, he spent 12 hours taking measurements from the cockpit of an airliner as it flew over the North Pole, and he trekked to northern Saskatchewan to capture the glow of the Northern Lights. He said he enjoyed getting immersed in these research projects at all levels, both in learning to operate the equipment and understanding the science.

Most of his time, however, was consumed with preparations for testing the SVS, the major objective of his flight. The system was developed by several researchers at the National Research Council, notably Lloyd Pinkney, who was the principal investigator. MacLean also got involved in developing the system, especially the advanced version, which incorporates laser scanning technology.

The prototype version, which was flown on his mission, features a computer system that rapidly analyzes video images from TV cameras and generates information about the location, orientation, and movements of objects, known as "payloads," that the astronauts are trying to grab with the Canadarm. (The term "payload" refers to satellites, instruments, and other scientific and technical equipment, as distinct from the space vehicle itself.) On the shuttle, the arm has handled many payloads of varying sizes and shapes, including satellites being released into orbit or captured for repair or return to earth, and experiment modules like the long-duration exposure facility (LDEF), a bus-sized container used for space materials-processing experiments.

Manipulators will also be used during construction of the space station, to handle and assemble the beams that make up the station's truss structure. This will involve complex manoeuvring at close quarters as the trusses are locked together. Then, when the station becomes operational, manipulators will be used to help dock the shuttle to the station, to load and unload the shuttle's cargo bay, and to haul supplies and equipment around the station. The sheer amount of work to be done, as well as the increasing size and complexity of the payloads being handled, makes it imperative to improve the efficiency of manipulator operations without compromising safety, and this is what the SVS is designed to do. Its accuracy and speed (it updates information 30 times per second) will allow astronauts to determine the proximity of the arm to the payload, and to the shuttle or the station itself, far more precisely than is possible using their eyesight (either looking out the window or viewing the camera images themselves).

The SVS is expected to be particularly helpful when, as often happens, visibility conditions are less than ideal. In space, bright, harsh lighting, deep shadows, and a lack of reference points often make it difficult for the human eye to accurately judge the position and movement of objects outside the shuttle. Moreover, visual cues can be suddenly and dramatically altered by lighting changes that occur as the shuttle moves from sunlight to darkness.

To help overcome these problems, a geometric pattern of dots, known as a "target cluster," is placed on the payload. The precise dimensions of this pattern (e.g., the distance between the dots) and its exact location on the payload are programmed into the SVS computer, which can then compare this information about the target cluster with the video images from the TV cameras and compute the changing position and orientation of the payload. For example, two dots in the cluster will appear to be closer together when the payload is closer to the shuttle than when it's farther away. Or if the payload is tilting upward, the dots will appear on the video images in a shape that is quite distinct from the rectangular pattern that appears when the payload is face-on to the camera.

The SVS rapidly translates these changes in the appearance of the target cluster on the video images into information about the payload's distance, speed, and orientation and displays that information in both numeric and graphic form on a TV monitor for the arm operator to use. The system can tell how fast the payload is moving in any combination of the six degrees of motion — up and down, right and left, back and forth, and rotating around any of the three axes: lateral (pitch), longitudinal (roll), or vertical (yaw). It can also accurately compute how far away the payload is.

In the tests performed on MacLean's mission, the target cluster was put on a specially designed metal plate, about the size of a piece of plywood, known as the Canadian target assembly (CTA). It resembled a large domino; one face was black with white dots and the other was white with black dots.

During the mission, the CTA was plucked from its resting place by the Canadarm and moved through a series of manoeuvres and then let loose to fly free. MacLean's job was to operate the SVS during these exercises and to determine how well it tracked the CTA and assisted the mission specialist controlling the arm. His objective was to do a thorough evaluation of the system's performance to help in designing operational versions for the shuttle and eventually the space station. Advances in remote manipulation technology may also have many applications on earth, in manufacturing, mining, and work in hazardous environments such as nuclear reactors and under water.

MacLean worked closely with the two astronauts responsible for arm operations, Lacy Veach and Tamara Jernigan. Before the flight, they'd spent a lot of time in the simulators, using

a ground-based version of the SVS to practise these procedures, including anticipating possible malfunctions and how
to work around them. The Canadian astronauts were allowed
to operate the arm in the simulators, although they are not
qualified to do so in space. The NASA astronauts also operated
the SVS system in these tests, "so we both understand the
other person's point of view," said Tryggvason. He said being
given the chance to operate the arm was an unexpected bonus
and entirely voluntary on the part of the NASA crew. Normally,
payload specialists are given instruction only on the systems
they'd need to use in space, such as housekeeping facilities
and the communications system and, for this mission, the
shuttle's closed-circuit TV system.

MacLean and Tryggvason were pleased with how well the
SVS performed in the preflight tests, which impressed the
NASA astronauts who saw it in operation. "It clearly gives the
arm operator a real advantage compared to what they have
now," Tryggvason said. "It really does improve their ability in
terms of accuracy and cutting the amount of time it takes to
do a task."

Training officer Parvez Kumar said that they were delighted
with "the feedback we're getting from NASA, particularly from
Steve's crew. They just shook their heads with disbelief, saying
that we're so far ahead of their own development in this thing
that they've kind of abandoned theirs and are using ours."

The news was equally good when the system was tried out
in space. Shortly after the successful completion of the flight,
Bruce Aikenhead said that a complete analysis of the videotapes would still be needed, but "looking at what we saw on TV
and listening to the air-to-ground comments, it . . . was apparent [the SVS] was working well. The guys on the ground who
were watching it were saying, "Look at that — piece of cake."

Several manoeuvres were done — most of them near the
end of the mission, so that the jiggling effect of moving the
Canadarm around would not upset microgravity materials-
processing experiments carried out early in the flight. After
MacLean checked out the SVS to see that it had survived the
rigours of launching, Lacy Veach used the arm to grab the
Canadian target assembly, lift it out of its berthing cradle in

the shuttle cargo bay, and put it back in again. The cradle was deliberately designed with less than two centimetres of clearance, so this exercise would simulate the degree of precision that will be required for several space station construction tasks, such as joining trusses or attaching the laboratory modules. Veach tried the exercise twice, once with the SVS and once without, and found it much easier to do with the help of the vision system.

MacLean said, "I pretended I was working with a big truss for space station, and I think the whole crew treated it like that, so when we did things according to plan, we knew that certain aspects of space station [construction] would be able to be accomplished. We can do the job; we have the accuracy to do the job."

Another test involved using the arm to move a payload along a triangular path on an inclined plane. "It went uphill and downhill and around the corner," said Aikenhead. Doing this without the SVS is a slow and tedious task, but it went much faster with the SVS, he said. "We saw [Veach] move slowly and gingerly initially, then he picked up speed and . . . he just zipped this thing around." This exercise simulated how the SVS might help the space station manipulator get from one point to another while avoiding collisions with structures and equipment on the station's beams, such as solar panels and radiators. "It's an option that could be developed in an operational [SVS] system for space station, where you could develop the path, put in the appropriate piece of software, and there's the display and away you go."

Finally, the CTA was released into free-flight, and the SVS was used to track its distance and orientation. Comparisons with readings from the shuttle's rendezvous radar (which is what the shuttle pilots depend on for close proximity operations with free-flying objects) demonstrated that the SVS was tracking the CTA with "very high accuracy," Aikenhead said.

Another test was intended to provide data that will be helpful during space station operations. It involved using the SVS to measure how firing the shuttle's thrusters affected the movement of the CTA. Doing this with an object of known size and mass will give engineers a baseline measurement that will be useful in analyzing how the station will be affected by shuttle thrusters firing in its vicinity. "It's a very good test for that," said MacLean. "You've got actual numbers on something real; then you extrapolate to the station and find out how accurate your model is." Even though the station will be much larger than the shuttle and the thrusters may have only comparatively small effects, this could still affect station operations and perhaps some scientific projects, especially materials-processing experiments, which are sensitive to minute gravitational effects. (The mobile servicing system will likely be used for the final stages of docking the shuttle, to reduce thruster firings so close to the station.)

The only problem MacLean encountered during the tests stemmed from "blooming," or washing out, of the camera images of the CTA, which occured when the light level was suddenly changed. "When the sun is setting, or if you get a reflection in one corner of the field of view, the iris tends to open and close, open and close, and it takes a while to settle down. The newer cameras don't do that," Maclean said."

Karl Doetsch, director of the Canadian space station program, said that the data on lighting problems gathered during MacLean's flight will be invaluable for the development of the space station system. "We'll certainly be looking at the experiment . . . to ensure that if we're relying on the camera [for] a certain operational task, that it's something that can be done reliably. And also finding out if there are some foibles, some conditions where it really didn't work, so that we can focus on solving those problems."

However, MacLean said, there's no camera in existence that can handle the problem if the sun is directly in the field of view — for example, behind the target and facing into the camera lens. (It's a problem not unlike that faced by any photographer who tries to get a good picture of someone when the camera is pointed at the sun.) The solution is to use a laser scanning device that can "see" the object even if the sun is in the field of view. "A good camera would make [the SVS] 90-percent operational; the laser makes it 100-percent operational," MacLean said.

The CTA was left in space to re-enter the earth's atmosphere and burn up. Because the amount of space junk collecting in earth orbit is a matter of growing concern, the shuttle was moved down to quite a low altitude specifically so that the CTA would burn up very soon after being released. The process took less time than expected because a solar flare caused the earth's atmosphere to expand, and "the atmosphere came up to meet it," Aikenhead said.

MacLean said this provided the shuttle crew with a memorable sight — both the CTA and the shuttle's tail were bathed in a green glow resulting from interactions with particles in the upper reaches of the atmosphere. "The atmosphere was thicker and more colourful [than usual], and the CTA probably was glowing more because of that." (He offered a less scientific theory as well, joking that the green glow might have had something to do with the fact that the CTA was released on Hallowe'en night.) "What a show — it was very artistic. That was neat because it allowed us to track it a little longer."

At one point, the CTA was about 365 metres below the shuttle, and it provided another dramatic image as it zipped along at about 25 times the speed of sound (Mach 25). "What that relative difference between us and the CTA does is give you a sense of its speed, so you get a true image of Mach 25, better than if you're just watching the tail of the orbiter cross over the clouds," MacLean said. "It was very impressive." The people on the ground were equally enthralled when they saw the video images sent down by the crew. "It was most astonishing," said Aikenhead. "There was scattered cloud over the ocean, they could see the cloud shadows, and there was the CTA going over the clouds at a tremendous speed. It looked like it had afterburners."

There were some tense moments during the mission as well — what MacLean referred to as "little bumps along the way." The worst occurred when he discovered that the laptop computer controlling the sun photospectrometer wasn't working. "The display screen got fried. We had tested it under that configuration on the ground and never had any problems."

After the ground support team did some remote diagnosis, they concluded "there was nothing we could do to save it, so there were some pretty gloomy people," said Aikenhead. "But the NASA people were astounding; they said, 'Don't give up. There are some other laptops on board; maybe you can reprogram them.'" A ground team lead by project manager Tom McElroy worked feverishly to rewrite the computer code for the NASA laptop and sent it up via the shuttle's data link to be loaded into one of the computers on board. MacLean worked doggedly to debug the code and get the new computer interfaced with the photospectrometer while keeping up with the rest of his duties. "He was hopping from one experiment to another, getting a few minutes with this and back to something else," Aikenhead said.

On the last day of the flight, NASA allowed the shuttle to move into the position required for MacLean to point the SPEAM instrument in the right direction. "I finally got the system working, [and] we were able to get the data we wanted," MacLean said. He was particularly pleased because the photospectrometer "is very sophisticated . . . better than any other instrument that exists," and it forms a crucial component of a world-wide network of instruments used to study changes in the earth's ozone layer.

MacLean had not specifically trained for such a repair job. "It was just a function of your experience," he said, adding that this was another example of the way human versatility in responding to unusual situations can rescue expensive and important space experiments.

For his trouble, he became the first shuttle astronaut to get a sunburn in space. The photospectrometer was designed to take measurements of a certain portion of the ultraviolet part of the spectrum, so NASA modified one of the shuttle's windows to allow ultraviolet light in. (Normally, special filters over the windows protect the astronauts from UV exposure.) "We provided a filter that would block the most harmful part of the ultraviolet, so the part of the UV spectrum they were looking for would get through," said Aikenhead. "The end result was that the exposure should be no worse than being at the beach. But nevertheless, Steve was cautioned to wear sunglasses. He had his Blue Jays cap on and his sunglasses, [but] he still got a sunburn on his cheeks."

With all his experiments, MacLean was putting in long hours. During a 20-minute live interview on CBC's *The Journal*, MacLean was asked to comment on a criticism that the mission was "a little light on purpose." He responded that "when I hear that, I think there must be another shuttle up here, because we're very busy. We're working a 16-hour day, we don't stop until it's time to go to bed, and for two or three days, I hardly got anything to eat because we were so busy."

In fact, in an unusual tribute that took MacLean by surprise, commander James Wetherbee commented that MacLean "is probably the busiest one of all of us. He has the most experiments and he's performed just superbly. It's obvious that Canada has sent us her best." At one point, Wetherbee asked mission control to allow one of the U.S. astronauts to take over some of MacLean's more routine tasks so he could devote more time to his experiments. The crew wanted to ease his burden because they were aware he still had some very complex SVS tests to perform near the end of the flight, according to Alan Mortimer, deputy manager of the CANEX experiments. He commented that Canadian astronauts are often more loaded down with work than their U.S. counterparts because they get far fewer chances to fly.

The demands of his experiments didn't leave MacLean much time for sightseeing, but he tried to steal as much time as he could to watch the earth go by, an endless source of fascination for the astronauts. Towards the end of the mission, he averaged only about four hours of sleep, partly because of the time it took to fix the SPEAM computer, but mainly because he took "a couple of hours each night out of my sleep . . . to look out the window. You could be up there for 10 days and never leave the window. There's just so much to see; each time you cross over something, if you're familiar with the area, it's fascinating to see. I loved it."

An astronomy buff, he was also fascinated by the show put on by the stars, the moon, and the planets. Much of the time, the shuttle was flying "upside down" in relation to the earth, so the earth appeared to be above them and the stars, below. "So here I am looking down at the stars. That in itself is neat because your whole life you look up and out at them, and here you are, floating above them." Allowing his eyes to get "dark-adapted" by turning off all the lights inside the shuttle provided some further astronomical delights. "At first, the universe looks like it normally would; in terms of the number of stars you see, it's like a clear night in a desert somewhere. But then after about 20 minutes, when your eyes dark-adapt, the stars start to push out at you, and the Milky Way seems to extend into the entire sphere that you're looking at. This was wild; I really got the impression that the earth was bathing in this milk bath. And [then] you start seeing the colours of the stars. [Some] were purple or orange or blue. I never expected that; usually you don't see that except through a telescope."

Having the earth above them provided a unique perspective of other events a little closer to home. "At the beginning of the mission, you had Venus coming out from underneath the earth — it was falling out from the earth, and then the waxing moon, the black moon, falls out after that, and then the sun [comes] up just a few minutes after the moon, so the moon is falling down into the full glory of sunrise."

He experienced many different sensations as the shuttle changed attitude. "Sometimes you feel like the shuttle is standing still and all this is rotating about you, and at other times, you sense that you're moving. If the shuttle flies nose down and you're looking out the pilot's window, you really get a sensation that you're falling around the earth." This was not a frightening sensation, like racing down from the top of a roller coaster, he said. "It seems slower than that — just sort of like you're faaaaalllling towards the earth. It's an illusion you get because of the attitude you're in and the way the windows are."

MacLean's mission did not get as much media coverage in Canada as Bondar's did, probably because it was competing for media attention with heavy hitters like the referendum on the Charlottetown constitutional accord, the U.S. presidential election, and the Toronto Blue Jays' World Series win.

MacLean did have some moments in the spotlight, during the unprecedented interview with *The Journal*, for example. After talking about the SVS experiment, MacLean provided Canadians with a tour of the shuttle's flight deck and mid-deck. A former competitive gymnast, he was also able to show off his microgravity gymnastics technique.

On the whole, however, the U.S. media seemed to be more interested in the flight, and particularly in the SVS experiment. "We got a lot of coverage," MacLean said later. "The American coverage was among the best they've ever had — at least that's what everyone was saying." One reason was that the crew did daily media briefings from space, something that had never been done before. NASA's own interest in the SVS was also a factor; its public affairs office "did an excellent job of showing what we were doing on the flight and what it meant for space station," said MacLean. "Each time we did a test with the SVS, we had the space station equivalent being shown on the ground behind it."

The media were also interested in the friendly rivalry among the crew members over the World Series. MacLean, a staunch Blue Jays fan, was asked by reporters on launch morning if he'd stayed up the night before to watch the game, the first ever played on Canadian astroturf. He admitted that he had — which prompted one CSA official to comment later that he was glad the game hadn't gone into extra innings. In space, despite his heavy workload, MacLean was able to follow the series. "The ground was really quite good — they pumped the scores up regularly. We had a lot of fun with it; we treated it as not just the Blue Jays against Atlanta, but as Canada against the United States. I will say that — I'll have to be diplomatic here — there was more than one Blue Jay fan up there." The morning after the Jays won the series, mission controller Bill McArthur wakened the crew by informing them that "there are celebrations in the streets of Toronto. Steve's going to collect on a few bets today." When reporters asked what MacLean had won from his American crewmates, Lacy Veach said that commander James Wetherbee had promised MacLean a free ride home. "I'd still be in orbit if they'd lost," MacLean joked.

He had carried a Blue Jays baseball and cap into space with him (along with crests of the Blue Jays, the Montreal Expos, and the Ottawa Senators), and he was tickled to see a cartoon that appeared in U.S. papers depicting the Atlanta Braves being retired by a baseball falling out of orbit. He commented that "people think I tossed a hat and a ball on at the last minute just to take advantage of [the World Series], but you have to put that stuff on five months in advance, so they had been in the bowels of the orbiter for a long time before we flew. I wish, instead of putting the ball on the orbiter early, I had put some money on the Blue Jays early."

Meanwhile, on the ground, the Canadian win nearly caused a second international flag incident in Houston. On the evening of October 24, a group of Canadians who were at the Johnson Space Center for the mission gathered in a hotel room to watch what would be the final game of the series. When the Jays emerged victorious, they skulked out and ran a Canadian flag up the flagpole in front of the space centre. As one of the perpetrators later recounted the tale, it took NASA's security people more than a day to even notice the Canadian flag, but when they did, it caused high anxiety. It was flying below the U.S. flag on the same pole, which is considered a major breach of diplomatic protocol, and NASA officials were worried that the incident might set off a furor similar to the one that had occurred a week earlier when the Canadian flag was carried upside down by the U.S. marines before a World Series game in Atlanta. But the CSA officials to whom the flag was returned offered straight-faced assurances to NASA that this would not be a problem.

When the mission was over, CSA officials were nearly hopping with satisfaction over how well it had gone. Mac Evans described it as a textbook flight and said that MacLean and Tryggvason made a very favourable impression on their NASA colleagues. "Everywhere I went at Houston and Cape Kennedy, they were singing the praises of Bjarni and Steve."

The SVS was also a hit. A few days after the flight, Evans said, "We have to analyze the data on the ground, but I talked to Lacy Veach when they came down last Sunday, and he's as strong, if not a stronger, supporter of SVS than he was before he flew. So this is not Steve saying it — it's the arm operator who flew the arm using the space vision system."

"His favourite expression up there was 'This is great. This is great,'" MacLean said. Veach later predicted that the SVS would revolutionize manipulator operations the way instrument flying had revolutionized aviation. Although MacLean thought this might be "stretching it a bit," because flying a plane is more complex than operating a manipulator, he agreed that operating manipulators would undoubtedly become more efficient, "and it really will be cheaper. They can save literally millions of dollars of simulator time on any given flight." This is because up to half the time spent training arm operators is devoted to techniques for moving the arm a single joint at a time, a slow and laborious backup method that would be used if the primary control system fails. The use of the vision system would make these single-joint manoeuvres far easier and thus greatly reduce the amount of preflight training, MacLean said.

With the success of the mission, NASA moved quickly ahead with plans to put the SVS on all the shuttle orbiters as well as on the space station. "NASA has already asked if our people can continue to support studies on an . . . operational SVS," said Aikenhead. Almost immediately after the flight, CSA started assembling a team of engineers to develop a hybrid system that could be used on both the shuttle and the space station. "There would be some features tailor-made to suit the shuttle and others for space station, [but] some degree of commonality is possible," Aikenhead said. "We'll start down that path as quickly as possible. There is some urgency to try and get an operational system into the shuttle."

"The bottom line is that NASA's knocking on our door," said MacLean. "When you consider that we're a foreign country, it's a major thing. It's a lot easier if you live in Texas to get stuff like that done. We really have a niche in this technology, [and] the reason is that we've got a solid team beneath us. Canada's good at this stuff."

As a matter of fact, an advanced version of the SVS is already available for testing — a legacy of the long post-*Challenger* hiatus, during which MacLean worked with engineers and researchers from the National Research Council, the Defence and Civil Institute for Environmental Medicine, and the University of Toronto to improve the system. (During that time, MacLean was also the astronaut advisor for machine vision research funded through the Strategic Technologies for Automation and Robotics (STEAR) program, which promotes earth-based applications of technologies developed by the space program.)

"Steve is integrating all the best ideas into a total state-of-the-art system, which I think will be a world-beater," said Doetsch. "We would look at it as being part of our evolution. We will have camera-based vision systems initially, and then we're going to upgrade, because there's no question in my mind that the vision system is very, very fertile ground for making significant advances which are very important terrestrially as well."

With his background in laser physics, MacLean's ambition is to promote the use of a laser scanning device that would enhance the ability of the SVS to track objects, particularly under poor lighting conditions. Two National Research Council engineers, Marc Rioux and François Blais, developed a workable laser system for the SVS, "so we have this prototype hardware system that can work against the sun, and now we're working on the software," said MacLean. "We modified the [laser] design quite a bit so that it would meet what we needed on space station. It really does the job. Everyone worked under the table and after hours to get it to the point where we could show people that it was a good idea." He admits it was a bit of an uphill battle at times: space station managers "have so much to do that when you say you can do something, they believe it when they see it. So that's what we did — we got it to where it worked, enough . . . to show them it was a good idea."

In fact, he was so enthusiastic about the system, despite the heavy workload associated with training for his flight, he could not pass up the opportunity while he was based in Houston to promote the advanced SVS system to NASA officials. "I've had 10 to 12 meetings already with shuttle management, and it looks very good. Maybe I shouldn't do it so much, but it's so exciting. We've got the advanced system down here, so the crew sees the potential of that system, as well as what we have for the flight. The flight will give us the data that will allow us to make the advanced system work that much better when we get to space station. It's the technology in the advanced system that's going to be used in the vision system on the station. There's just a world of difference in capability. The present system can only track four targets at a time; the advanced system can track up to 200. The bottom line is that they cannot do certain portions of the space station construction without it — we showed that while we were down there while I was training."

MacLean and Tryggvason were back at work on the SVS almost immediately after MacLean's flight. They participated in evaluation tests of the advanced SVS with NASA engineers and astronauts at the Johnson Space Center, during which the system was used to simulate a variety of space station construction tasks. These activities are expected to lead to another shuttle flight, during which the prototype of the advanced system will be tested by a Canadian astronaut. Aikenhead said the plan is to fly "a member of the design team who is already trained as a payload specialist and who could then continue to work with that team." Bjarni Tryggvason fits this profile perfectly, and he's made no secret of his interest in flying the mission. He's "a logical choice," Aikenhead said. "We'll have Bjarni as a member of the team staying on down there in the continuing evaluation of the SVS for space station assembly."

As for MacLean, he plans to stay in the astronaut program. "We've got some exciting days coming up. I think the flight rate is going to increase [for Canadians], and I'm going to work on the next step on this project." He hopes to get another flight — "I'm young enough for that. [But] we'd be pretty shallow if the only reason we were in it was for the flight. Hopefully we'll get this technology onto station and then we can really do some interesting things with it. It's a long process, and it's a very exciting process, and I feel I'm in a good position to contribute. All those doors are open to me, [so] right now, my plan is to stick around. It's fun to work on this stuff."

6

On the Road to *Freedom*

The international community is going to want to see a fair amount of experience with international crews working together before they actually do it on space station.
MAC EVANS, VICE-PRESIDENT, OPERATIONS, CANADIAN SPACE AGENCY

If things had gone according to plan, the dream of an earth-orbiting, permanently manned international space station would be approaching reality by now. In January 1984, then U.S. president Ronald Reagan officially gave NASA the mandate to build space station *Freedom*, instructing the agency to "do it within a decade." Though his charge was reminiscent of the one given by John F. Kennedy that took U.S. astronauts to the moon in 1969, the results were not the same; this time, NASA and its international partners, Canada, Europe, and Japan, find themselves in a struggle to get the station built and occupied by the turn of the century.

Their ambitions have been tempered by economic and political realities: like the space station itself, the dream has been downsized and restructured. It will take longer to build, the crews will be smaller, the missions shorter, the quarters more cramped, and the costs greater. Nor will opposition to the project disappear; in fact, it may intensify if the global economy continues to stagnate. Yet the international partners appear to be steadfast in their commitment to making it happen eventually. Although NASA's budget comes under yearly as-

144

sault in the U.S. Congress, when it comes right down to it, "when you look at the votes, it's consistently strongly in favour of space station," says Mac Evans. "I take that as a very strong political commitment on the part of the United States to continue."

However, with the election of Bill Clinton as president in November 1992, and changes in the make-up of Congress, how space station funding will fare in the future remains to be seen. During the election campaign, vice-president Al Gore, who subsequently became chairman of the U.S. National Space Council, said the Clinton administration would complete the space station program, but there's no way of telling what might happen when the postelection realities of running the United States hit home. And space station opponents are clearly hoping that the influx of new blood into Congress will tip the scales in their favour. Congressman Tim Roemer, for example, predicted that newly elected members would be "more receptive to making tough decisions like cancelling the space station" and said that getting enough votes to do so would be "very achievable."

If this comes to pass, the role of the international partners may become more important than ever. NASA originally invited them into the program largely to provide ballast during political storms, and they've already proven to be up to the task. In the spring of 1991, for example, when Congress was making what appeared to be serious noises about savaging the space station budget, representatives of all three partners rushed to Washington to give the U.S. government a collective piece of their minds for threatening a program to which they had already committed large amounts of their own space budgets.

"The presidents of the three space agencies issued a joint communiqué," said Evans, who was among the Canadian contingent that went to Washington. "We wanted to make sure that in making their decisions, they understood the implications of space station to the international partners. We noted that this program was initiated by the United States, [and that] the international partners were invited by the president of the United States to participate in the program. All three of us, the Europeans, the Japanese and ourselves, thought it was a good opportunity and in essence restructured our whole space programs around space station. It's a significant element of the space programs of all the international partners."

The communiqué noted that the program was also "a test of the ability of the participating nations to concert their efforts towards a common goal. The withdrawal of the United States from this program would have far-reaching implications for the future of international cooperation in space."

This theme was picked up in a letter to the space committee of the U.S. House of Representatives from Derek Burney, then the Canadian ambassador to the U.S., who said the space station program was unique because "it tangibly demonstrates the willingness and commitment of the nations involved to ensure that space must be developed in ways that will benefit all humanity. The United States exercised leadership when it invited other nations to join it . . . and the action sent a watershed signal to the rest of the world . . . For the United States to withdraw at this point . . . would at the very least indicate that nations have failed in this important attempt to work together for the peaceful development of space and will profoundly affect the prospects for international collaboration in this area."

In the end, the project received a substantial majority of votes in Congress. Evans said "it was a clear indication that they recognized two things — the importance of space station to their own program, and since the international aspect played a large role in the debate, I would say they recognized the importance of space station to the international partners and the importance of the international partners to the U.S. program."

As for the astronauts, *Freedom* is, to a large extent, their *raison d'être*. The shuttle was never intended as, nor equipped to be, a full-time research lab and crew base. As its official name — the Space Transportation System — suggests, it was meant to get people there and back. But, in the absence of a space station, there is no "there," so they've been, in Roberta Bondar's words, "making do" with the shuttle. Still, shuttle flights are playing an important role in preparing the international astronaut corps for the space station era. Not only do they provide valuable experience in space operations per se, but they also allow astronauts, as well as ground-based scientists and engineers from different countries, to work together. (Bondar's IML mission was a model of this kind of collaboration.) Canadian space officials are convinced that it will be difficult for any astronaut to be accepted as a member of the space station crew without some previous space flight experience.

Thus, by the end of 1992, with Bondar's and MacLean's missions successfully completed, CSA officials turned their attention to negotiating future flight opportunities. In their minds, Canada has two flight commitments stemming from the agreements with NASA on space station participation and mission specialist training. The space station agreement states that a Canadian astronaut will participate in the on-orbit verification of the mobile servicing system, and the mission specialist agreement says that at least one of the two Canadians sent for training in 1992 will be assigned to a shuttle flight. CSA officials do not expect that the two commitments will be rolled into only one flight for a Canadian astronaut, notwithstanding the fact that a Canadian mission specialist will probably be required for the MSS verification flight. "We think those are separate things," said Evans. "In our minds, the commitment to have a mission specialist fly from the Garneau/Hadfield pair is different from the commitment to have Canadian astronauts participate in the checkout of our equipment. My guess is that there will be a Canadian mission specialist that flies in advance of [the MSS flight]."

Moreover, CSA officials were hoping to persuade NASA to let *two* Canadians fly MSS verification flights because of changes to the launch schedule that occurred as a result of the restructuring of the space station program. Originally, all of the MSS components were to be launched on a single flight, but when the dust settled from the restructuring, it was decided that the major elements of the Canadian technology would be carried up on two flights instead: the large arm (the space station remote manipulator system) in the last half of 1996 and the smaller special purpose dextrous manipulator, along other equipment, about a year later. Since both will require in-flight verification, "we're trying to turn it into two flights [for Canadian astronauts]," said Karl Doetsch, director of the Canadian space station program.

If this is approved, Canadians will have at least three flight opportunities before 1997 — and possibly four if both Garneau and Hadfield receive flight assignments before then. It's even possible that a second set of Canadian mission specialists could be trained for the MSS verification flight or flights.

Another, earlier, flight opportunity might arise from the success of the SVS experiment on Steve MacLean's mission. Evans thinks it would be feasible to fly a mission, perhaps in late 1994 or early 1995, to test the prototype of an operational advanced space vision system, possibly in an exercise simulating construction of the space station trusses by the Canadarm. If NASA plans to use the SVS for the real construction work, Evans predicts "they're going to want to try it out. So I think there's a strong potential for a future flight of the SVS system in more of an operational configuration, maybe with more bells and whistles and maybe even with a demonstration of [truss] joining." And since NASA's so interested in the technology itself, there's even a possibility it might be willing to foot part of the bill for the mission.

Evans believes this mission would be open to a payload specialist. "I don't think we would have any problem demonstrating that it would take extraordinary training for their people to come up to the same level that Steve and Bjarni were at. I think that for the next SVS flight, we can meet all the conditions that NASA requires for a payload specialist to fly."

There will be additional flight opportunities between 1997 and the start of permanently manned operations aboard the space station after the turn of the century. These will be "hybrid" shuttle/station flights throughout what's known as the station's "astronaut-tended" phase, during which the shuttle will remain attached for missions lasting about two weeks, and astronauts will work aboard the station while using the shuttle as their living quarters. Canadian astronauts should be eligible for these flights. When the station progresses to permanently manned status, crews will remain on board continuously, rotating every three to six months, and the shuttle will assume the role for which it was originally intended — ferrying astronauts, equipment, and supplies between earth and the station.

With all these opportunities, "I think we're going to find there are a lot of Canadians flying," Evans concluded. There's a fairly high probability of perhaps three to six shuttle flights for Canadian astronauts before the turn of the century and possibly even more. And there may be other opportunities: Canadian officials have not given up on the idea of getting a Canadian on board the Russian *Mir* station. In the late 1980s, Marc Garneau had invested some time exploring that possbility and had been enthusiastic about the potential opportunities. With *glasnost* in full swing, the Russian space program was opening up to Westerners more than ever before. In 1987, while attending a space conference in Moscow, Garneau was among a group of people invited to visit Star City, where Soviet cosmonauts are based. They met about 30 to 40 cosmonauts and were given a tour of the training facilities, including the large water tank where cosmonauts were practising EVA procedures with a full-scale mockup of the *Mir* space station. It was a remarkable event because "at the time, it was a closed-off place," he said.

Garneau strongly pushed the "Soviet option" with the CSA, and by early 1990, he believed it had come to be viewed as a real alternative. "It's a step forward, in the sense that we're not just saying that we can only fly on the shuttle," he said at the time. "I think senior management is going to be looking at all options for flights beyond Steve's and Roberta's. And the Soviets have informally provided enough encouraging signs that we believe that if we were to go into negotiations with them, there would be very little standing in the way of organizing a mission."

But any question of Canadian participation was put on indefinite hold after the 1991 collapse of the Soviet government and the subsequent social and political upheaval that left their space program literally up in the air. (One cosmonaut was actually stranded aboard the *Mir* station for months before the new Commonwealth of Independent States got its act together enough to bring him back.)

However, although the new Russia continued to struggle with severe financial problems, by late 1992, the message it was sending out to the rest of the world was that its space program was still open for business. CSA president Roland Doré said that at the World Space Congress in September 1992, "it was clear . . . that Russia will continue to be a major force in space. Both the Russian Space Agency and the Russian academies received considerable attention from delegates from other nations. Everyone is interested in working with Russia, and it is clear that the end of the cold war means major new opportunities for the space program of the West." Indeed, that fall, Russia and the United States struck a historic deal that will culminate in shuttle flight for a cosmonaut, scheduled for November 1993, and a 90-day *Mir* assignment for a U.S. astronaut. The agreement also calls for the shuttle to visit the *Mir* station; it will carry up two replacement cosmonauts and return two others to earth.

A *Mir* flight might prove to be particularly attractive as a life sciences mission, to counterbalance the strong emphasis on engineering and technology development that will probably dominate most of the shuttle missions flown by Canadians before the turn of the century. The Russians, of course, still hold the record for long-duration space flight, and in recent years, they've become much more interested in collaborating with other countries in space physiology research. Since 1989, Canada and Russia have worked together on a number of joint science projects focussing on the long-term biological and physiological effects of space flight. In 1989, a University of Toronto rat experiment flew on the Russian biological satellite *Biocosmos*, and in 1992, several radiation-detection devices were transported to the *Mir* station to begin a year of collecting data on the radiation levels in space.

As humans stay for longer periods of time in space, the cumulative effects of radiation exposure will become increasingly significant. "The Russians have done a lot of research on the effects of long-duration flight," said Alan Mortimer, head of the CSA's life sciences program. "Our collaboration with them should be extremely beneficial as we prepare for longer-term presence of humans on space station *Freedom* and potentially beyond, to the moon and Mars." The research may also improve radiation devices used on earth. "Radiation detection is important for the well-being of humans both on earth and in space," said Ron Wilkinson, manager of CSA's User Development Program, which funded the *Mir* experiments. "This technology has had tremendous commercial success in terrestrial applications, and only now are we beginning to realize the potential for space applications."

There appears to be only one real obstacle now standing in the way of flying a Canadian astronaut on *Mir*. "It's merely a matter of money," said one CSA official, his wry tone conveying the hard reality that, in these cash-strapped times, funding is typically a greater problem than technical challenges or political barriers. And, as expensive as astronaut training is, it's not all that's involved: funding must also be found to support scientific experiments for the mission, including several years of ground-based preparatory research and the development and testing of equipment. A *Mir* mission would cost $20- to $35-million (U.S.), depending on its duration.

Doré said that, in order to "strike a new balance," Canada is also interested in exploring joint ventures with the European Space Agency (ESA). In recent years, however, ESA has been forced by economic and political realities into a serious overhaul and retrenching of its space plans. Gone from the agenda, at least for the time being, are the *Hermes* manned space plane (a kind of mini-shuttle) and a free-flying, astronaut-tended space laboratory. Moreover, the *Columbus* manned laboratory module that ESA is building as its contribution to *Freedom* has been cut in size, and its schedule has been stretched. This project could face continuing political troubles in years to come; ESA recently decided to review the program in early 1995.

As an associate member of ESA, Canada was invited to submit applicants for its first group of astronauts in 1991, which would have made Canadians eligible to fly as ESA-sponsored space station astronauts or as crew members aboard *Hermes*. "They were quite interested in having us supply candidates, [but] we just weren't ready," said Evans. At the time, he believed there might be future opportunities, but ESA's reduced circumstances makes this uncertain.

As for Japan, Doré said Canada will try to work more closely with their National Space Development Agency, but it "seems less open to the prospects for joint ventures." However, there may be opportunities for scientific collaboration with the Japanese Institute of Space and Astronautical Science.

For now, most of the energies of the world's space-faring nations are concentrated on establishing a permanent human presence in low earth orbit. But dreams and schemes for venturing farther afield — back to the moon and on to Mars — are already being touted. And Canada's two major areas of space research — robotics and human physiology — fit in very nicely with the way things are evolving, for both are areas of research essential to the success of long-duration and long-distance space flight.

The partnership between humans and robots — using the strengths of each to offset the weaknesses of the other — is growing rapidly in space because of the need to minimize risks and cut the costs of operating in such a hostile environment. What happened with the space station was a case in point. The original plan for constructing the station in space, which required astronauts to assemble the trusses by hand, had to be revised after NASA realized that a dauntingly large amount of EVA time would be required.

To reduce the cost and the risk to astronauts, a new plan was devised that involved pre-assembling sections of the trusses on the ground and using robotic manipulators to link them together in space. This plan greatly enhanced the role of the Canadian-built manipulators, both the shuttle's Canadarm and the space station's MSS. "My own view is that both of them are going to be used extensively throughout the assembly of the station" said Doetsch. "The Canadarm on the shuttle is going to get more use than had originally been planned because it's going to be the plug-in unit." In the early stages of assembly, the shuttle will bring up one truss section, the Canadarm will put it in place, then the shuttle will go back for another section. "The shuttle will move to berth with the component of the station that's up there — it will be berthed using the Canadarm and locked in — and then

Canadarm will take out the next component, which is in the bay, pick it up, and plug it in." At some point, the mobile servicing system will be positioned on the growing length of truss, and it will be used for constructing sections beyond the reach of the Canadarm.

This plan is expected to reduce the amount of EVA time to manageable proportions, Doetsch said. "So the robots are essential. The combination of robots and crew now no longer have EVA as a bottleneck in the assembly sequence, and I think that's a very important step forward."

At the time the plan was announced, Evans commented, "For the first time, we believe the assembly sequence will work. We always felt that erection and assembly of trusses in space were very complicated, and no one had demonstrated to us that the [EVA] concepts were going to work. We kept telling them so." He added that "the role of robotics is now enhanced over what it was" and predicted that the assembly of the space station is going to be "quite a Canadian robotic show."

This collaboration between humans and robots will continue and become even more essential as humanity moves farther out into the universe. But the other part of that equation — the human body — also requires further study, for it will be operating in an environment completely different from the one in which it evolved for millions of years. Humans have only been in space for about three decades, mostly for relatively brief periods of time, and we still know very little about the effects of long-duration flights on human physiology. Bob Thirsk has commented that some of the most challenging issues facing space engineers in the future will involve human factors and life support — "the medical side, the psychological and motivational aspects of maintaining human performance in space, as well as the engineering aspects." The space program is "going to be a big employer of engineers with life science interests."

These interests will range all the way from providing urgent medical care on the space station to keeping people alive and well on a three-year mission to Mars. With his background in emergency medicine, Canadian astronaut and physician

Dave Williams hopes to make a contribution to the development of laser surgery techniques that would permit bloodless
operations to be performed in microgravity. Ken Money,
meanwhile, has been a member of an international scientific
group, the International Academy of Astronautics, which has
been studying the long-term human factors and health issues
associated with a trip to Mars. These issues include radiation
exposure, the psychological effects of isolation, confinement
and perception of danger, and the physiological effects of
microgravity.

"Radiation could be a show stopper," Money said. "It isn't
actually known exactly what the radiation flux is that far beyond earth, and especially, it's not known how that will interact
with humans." And it's not just a matter of worrying about the
normal levels of radiation out there; a huge flare on the sun
could send a lethal wave of radiation washing over a spacecraft
on its way to Mars. Nor is the solution a simple matter of
providing radiation shielding; not only is such shielding heavy
and therefore expensive, it can actually make the radiation
environment inside a spacecraft worse under certain circumstances. Cosmic radiation, the kind coming in from beyond
the solar system, moves very fast and the particles slice
through the thin walls of the spacecraft quickly, but when
these particles hit a heavy wall of shielding, they can divide
into a shower of slower-moving secondary particles that are
more damaging to biological tissues than the primary
particles. This creates a curious situation in which astronauts
should stay inside a heavily shielded shelter during a solar
flare to protect them from solar radiation, but stay out of it
during normal periods because secondary particles from cosmic radiation would raise radiation inside the shelter to higher
levels than those outside.

Even with a shelter, astronauts going to Mars may well
receive higher doses of radiation than would be acceptable in
an industrial setting on earth. As a result, Money predicts that
older astronauts will be sent on such missions. One reason is
to avoid damage to people of childbearing age. "The females
should be postmenopausal because the damage to subsequent generations from radiation could be very serious."

The other reason — a hard but realistic one — is that age could be a factor in reducing the impact of any cancers caused by the radiation exposure. It might take cancers 10 to 20 years to develop, and older astronauts are statistically more likely than younger ones to die of other causes before that. Money says Mars crews may well be 55 or older. "Why send young guys who might get cancer? You have to double your shielding for them. Send old guys and their wives or old wives and their husbands." (There have been several married couples among NASA's corps of astronauts, and one couple flew together for the first time in 1992. If there are no suitable candidates already in the program, NASA "might have to look ahead and recruit them, that's all," Money said.)

This raises an issue that is probably NASA's least favourite — sex in space. Whenever NASA officials are asked about the subject — and they have been frequently since women started flying — they usually dodge it by pointing out that there's "just no privacy" on the shuttle. They even get squeamish about scientific experiments on related issues, according to one scientist, Lynn Wiley, an expert on mammalian development, who says she's never been able to convince NASA to allow the collection of in-flight sperm samples to learn more about the effects of microgravity on the human reproductive system. Fortunately, some relevant information can be obtained from blood and urine samples, and Wiley allows that semen tests on shuttle flights might be onerous and distracting.

Money says, however, that there will be no way to avoid the issue on a Mars mission. "I keep telling them that if you don't make provision for sexual function, you are out of your mind. I keep saying, you should send only married couples — of course, I am very old-fashioned — or at least send people in a stable sexual relationship. But don't send up three men and three women for three years without having settled the matter, or you could very well end up with people competing for sexual partners during a flight. Now that's not to be tolerated; it's a risk we don't have to take." Serious sociological and psychological problems could result from this abnormal situation, and "to send married couples would make it much more normal."

In fact, making space "more normal" may be the key to

protecting the psychological health of astronauts on long-term flights. As mission duration increases and the stress of living with a small group of people in a hazardous, closed environment mounts, so does the potential for psychological and social problems. B. J. Bluth, a NASA engineer who has studied sociological issues related to long-duration space flight, notes in one scientific paper that both U.S. astronauts and Russian cosmonauts have "lost data, ruined experiments, lost equipment, and manifested a long list of symptoms . . . [including] fatigue, sleep disorders, irritability, depression, anxiety, mood fluctuation, hostility, social withdrawal, vacillating motivation, boredom, tension, and lowered efficiency." It is worth noting that this has occurred in crews consisting entirely of men with similar backgrounds. If you stir into the pot the effects of male-female dynamics and the presence of different races and diverse cultures — as will be the case on space station and other long-duration flights — you have a potentially explosive mixture.

Russian cosmonaut Valery Ryumin, who flew two six-month missions, once noted that "all one needs to effect a murder is [to] lock two men into a cabin, 18 feet by 20, and keep them there for two months." He described the "deeply human" problem of adapting to living in space with another person. "We have to solve [our problems] together, taking into account the feelings of the other. . . . We are totally alone. Each uttered word assumes added importance. One must bear in mind, constantly, the other's good and bad sides, anticipate his thinking, [and] the ramifications of a wrong utterance blown out of proportion."

"There is no place you can go to blow off steam," said Bluth. "You are always in the same place, always with the same people. In larger groups, you can find people that you are in sync with, but in very small groups, the chemistry of the group makes a big difference." This is a major reason why space station crews will train together extensively before their flights. There's more than technical competence at issue here: the crews must knit together as people as well as workers. "The probability of having somebody up there whom nobody else has ever met before is low."

In the Soviet program, which has devoted far more serious attention to psychological issues than NASA has until recently, cosmonaut crews were once sent on cross-country car trips across Siberia. "They figured that was one good way to find out whether they could live in isolation and confinement," said Bluth. Subsequently, however, the Russians developed a much more sophisticated program of stress training and psychological and compatibility testing. And once the crews were in space, they were monitored by a psychological support team on the ground that observed the crew's activities on television. Bluth said they would watch for subtle signs of stress, such as how far the crew members would stay away from one another. "The social-distance information can reveal unspoken tensions."

Social scientists are particularly concerned with what happens after the first month in space, when the initial excitement and novelty have worn off and an abrupt deterioration in the crew's mood, motivation, and performance can occur. To help counteract this effect, the Russians provided their crews with many psychological boosters, permitting and encouraging direct contact with family, friends, and favourite celebrities on the ground. There were frequent two-way video conferences, and cosmonauts received letters, gifts, and videotapes from home. One, who got a tape of his daughter's birthday party, said, "I would put it on when I felt particularly homesick. You . . . get engrossed in it, and it seems like you're with your family." When another cosmonaut was sent his much-missed guitar, he started organizing sing-alongs with the people on the ground. "The funny thing is the guitar floats, so it is not that easy to play," noted Bluth.

Even with all this, cosmonauts did get tired, irritable, and fed up with themselves, with each other, and with the ground controllers. One of them wrote in his diary, "The most difficult thing about this flight is keeping calm in dealing with mission control and with the other crew members because pent-up fatigue could generate serious friction. I should work on cultivating great patience." Bluth said there was at least one case where a cosmonaut "underwent a sudden personality change and became irritable and troublesome. He began complaining about the organization of work on board and his dissatisfaction with the difficult living conditions." His behaviour was attributed to excessive consumption of sleeping pills because he had been unable to rest.

One of the keys to successful long-duration space flights is providing comforts and amenities people are used to on earth — good food, privacy, quiet, cleanliness, and leisure time. Space engineers are already working with psychologists and human factors experts to design kitchens, bedrooms, showers, and work spaces that will help rather than hinder the crew's adjustment to living in microgravity, and they are considering the psychological impact of colours, textures, music, and other sounds. The goal of these "habitability" experts is to make a small spacecraft as much of a home-away-from-home as possible.

According to a former NASA doctor, Patricia Santy, social and psychological factors will make the difference between success and failure in colonizing space. The space agencies of the Western world have come to recognize this reality and are now putting considerable effort into psychologically pre-screening astronauts. But no amount of screening can completely eliminate the risk of psychiatric disorders occuring in space, and it may be necessary to deal with psychoses, as well as depression and anxiety, even among such highly trained people. In fact, she suggested the possibility of an entirely new anxiety disorder — fear of space — that would be a combination of a fear of flying and agoraphobia. Therefore the space station must be equipped to restrain and sedate individuals who may suffer from paranoid or psychotic reactions. "Management of the psychotic state will have to be medically aggressive. Due to a lack of judgement and insight, which are often symptoms of the psychotic process, these individuals may place their crewmates at significant risk for disaster."

Bluth said that severe psychiatric emergencies are likely to be rare because "the population [of astronauts] is so highly selected that the chance of having outlandish things happen is not high." Still, such things do happen: Antarctic science bases and nuclear submarines also have highly selected populations, and some of their people have experienced mental disturbance and psychiatric disorders.

Despite these risks, and whether we're ready for it or not, it appears that humanity is heading out into space. "If we waited until we had all the problems solved, we'd never go," said NASA doctor James Logan. Like many others involved in the space program, he believes that the exodus of the human species is inevitable. We can only speculate on the implications for our evolution — on how this exodus will transform us, not only technologically and culturally, but also, eventually, psychologically and physiologically. What of future generations of people who will be born in space? It may be that humans of the future will have a choice between living with gravity or living without it. Would someone who has never set foot on earth be able to walk upright or balance properly? Would such people, born to float freely in microgravity, even have the slightest interest in contending with the miseries of earth's gravity?

We have pondered these questions for many years in science fiction. Now we are pondering them for real.

Index